All facts, no fiction with CGP!

Quick question — do you own CGP's Knowledge Organiser for Higher GCSE Maths?

You do? Excellent! Now you can use this Knowledge Retriever to check you've really learned everything.

There are two memory tests for each topic, plus mixed quiz questions to make sure that everything has properly sunk in. Enjoy.

CGP — still the best! ☺

Our sole aim here at CGP is to produce the highest quality books — carefully written, immaculately presented and dangerously close to being funny.

Then we work our socks off to get them out to you — at the cheapest possible prices.

Contents

Section 6 — Pythagoras and Trigonometry

Section 7 — Probability and Statistics

Published by CGP.
From original material by Richard Parsons.

Editors: Sarah George, Sharon Keeley-Holden, Samuel Mann, Sean McParland, Caley Simpson.

With thanks to Glenn Rogers for the proofreading.
With thanks to Emily Smith for the copyright research.

Printed by Elanders Ltd, Newcastle upon Tyne.
Clipart from Corel®

How to Use This Book

Every page in this book matches a page in the Higher GCSE Maths **Knowledge Organiser**. Before using this book, try to **memorise** everything on a Knowledge Organiser page. Then follow these **seven steps** to see how much knowledge you're able to retrieve...

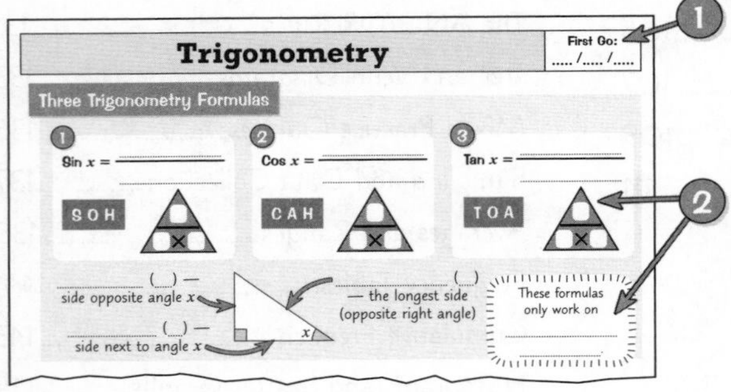

1. In this book, there are two versions of each page. Find the **'First Go'** of the page you've tried to memorise, and write the **date** at the top.

2. Use what you've learned from the Knowledge Organiser to **fill in** any dotted lines or white spaces.

   You may need to draw, complete or add labels to tables, graphs and diagrams too.

3. Use the Knowledge Organiser to **check your work**. Use a **different coloured pen** to write in anything you missed or that wasn't quite right. This lets you see clearly what you **know** and what you **don't know**.

4. After doing the First Go page, **wait a few days**. This is important because **spacing out** your retrieval practice helps you to remember things better.

5. Now do the **Second Go** page.

   The Second Go page is harder — it has more things missing.

6. Again, check your work against the Knowledge Organiser and **correct it** with a different coloured pen. You should see some **improvement** between your first and second go.

7. **Wait** another few days, then try to recreate any methods, formulas, tables or diagrams from the Knowledge Organiser page on a **blank piece of paper**. You can also have a go at any **example questions**. If you can do all this, you'll know you've **really learned it**.

There are also **Mixed Practice Quizzes** dotted throughout the book:

- The quizzes come in sets of four. They test a mix of content from the previous few pages.
- Do each quiz on a different day — write the date you do each one at the top of the quiz.
- Tick the questions you get right and record your score in the box at the end.

Types of Number and BODMAS

First Go: ..... /..... /.....

Seven Types of Numbers

		Definition	Examples
1		Whole number	–16, 0, 2, 547
2	RATIONAL	Can be written as a	$0\left(=\frac{0}{1}\right)$, $0.44...\left(=\frac{4}{9}\right)$
3	IRRATIONAL	Can't be written as a — never-ending, non-repeating	$\sqrt{2}$, $5\sqrt{3}$, π
4	NEGATIVE		–21, –3.6, –0.01
5		In a number's times table (or beyond)	Of 3: 3, 6, 15, 42
6		Divides into a number	Of 10: 1, 2, 5, 10
7	PRIME	Only factors are and	2, 3, 17, 43

1 is NOT ....

Two Rules for Dealing with Negative Numbers

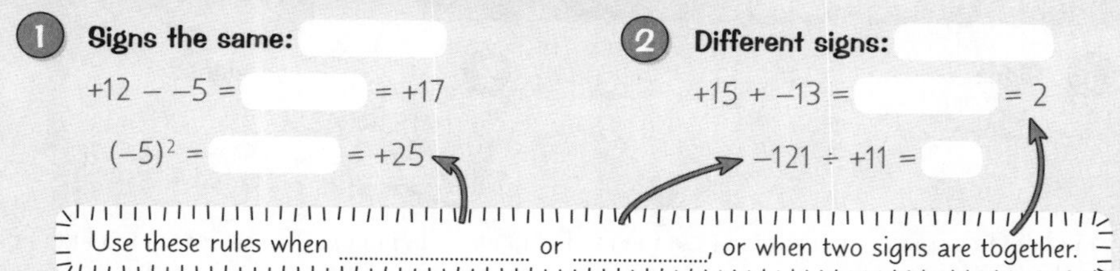

1. Signs the same:

$+12 - -5 = \quad = +17$

$(-5)^2 = \quad = +25$

2. Different signs:

$+15 + -13 = \quad = 2$

$-121 \div +11 =$

Use these rules when .... or ...., or when two signs are together.

BODMAS

BODMAS gives the order of operations:

1.
2. Other
3. Division and
4. and Subtraction

'Other' is things like ....

EXAMPLE

Find the value of $9 - (3 + 1)^2 \times 2 + 5$.

$9 - (3 + 1)^2 \times 2 + 5$

1. $= 9 - \quad ^2 \times 2 + 5$
2. $= 9 - \quad \times 2 + 5$
3. $= 9 - \quad + 5$
4. $=$

$=$

Work left to right when there's only addition and subtraction.

Second Go: ...... /...... /......

Types of Number and BODMAS

Seven Types of Numbers

	Definition	Examples
1		–16, 0, 2, 547
2	Can be	$0 \left(=\frac{0}{1}\right)$, $0.44... \left(=\frac{4}{9}\right)$
3	Can't be	$\sqrt{2}$, $5\sqrt{3}$, π
4		–21, –3.6, –0.01
5		Of 3:
6		Of 10:
7		2, 3, 17, 43

1 is ....

Two Rules for Dealing with Negative Numbers

1. Signs ______ : ______

   +12 – –5 = ______ = ______

   $(-5)^2$ = ______ = ______

2. ______ signs: ______

   +15 + –13 = ______ = ______

   –121 ÷ +11 = ______

Use these rules when .... or ...., or when ....

BODMAS

BODMAS gives the ______ :

1.
2.
3. and
4. and

.... is things like ....

EXAMPLE

Find the value of $9 - (3 + 1)^2 \times 2 + 5$.

$9 - (3 + 1)^2 \times 2 + 5$

1. =
2. =
3. =
4. =

=

Work left to right when there's only addition and subtraction.

Multiples and Factors

First Go: ...../...../.....

Four Steps to Find Factors

1. List factors ________, starting with 1 × the number, then 2 ×, etc.
2. Cross out pairs that don't ________.
3. Stop when a number is ________.
4. Write factors out clearly.

EXAMPLE

Find all the factors of 20.

1. 1 × ____
   2 × ____
2. ~~3 ×~~
   4 × ____
3. 5 × ____
4. So the factors of 20 are: ________

Finding Prime Factors

PRIME FACTORISATION — writing a number as its ________ multiplied together.

Three steps to use a Factor Tree:

1. Put the number at the top and ________.
2. ________ each prime.
3. When only ________ are left, write them in order.

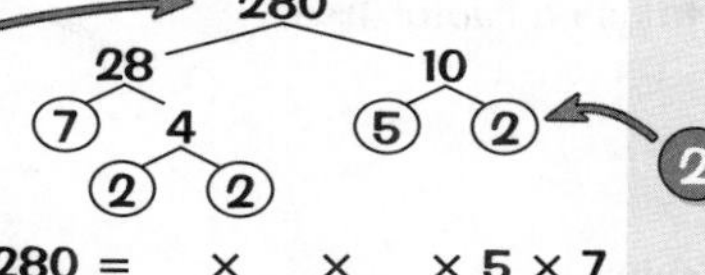

280 = ... × ..... × ..... × 5 × 7
= .... × 5 × 7

Lowest Common Multiple (LCM)

LCM — the ________ number that divides ____ all numbers in question.

Find it from ________ factors in two steps:

1. List all ________ factors in either number.
2. ________ together.

EXAMPLE

Find the LCM of 8 and 14.

8 = 2 × 2 × 2
14 = 2 × 7

1. 2, 2, 2, 7
2. .... = ...

If a factor appears more than once in any number, list it that many times.

Highest Common Factor (HCF)

HCF — the ________ number that divides ________ all numbers in question.

Find it from ________ factors in two steps:

1. List all ________ that are in ________ numbers.
2. ________ together.

EXAMPLE

Find the HCF of 36 and 90.

36 = 2 × 2 × 3 × 3
90 = 2 × 3 × 3 × 5

1. 2, 3, 3
2. .... = ...

You can also find LCM/HCF by listing the .... of both numbers and taking the smallest/biggest number that appears in ....

Second Go:
...../...../.....

Multiples and Factors

Four Steps to Find Factors

1. List , starting with
, etc.
2. Cross out pairs .
3. Stop .
4. Write .

EXAMPLE

Find all the factors of 20.

1.
2. ~~3 ×~~
3.
4. So the factors of 20 are:

Finding Prime Factors

PRIME FACTORISATION — writing a number as its .

Three steps to use a Factor Tree:

1. Put
2. Circle
3. When only

280

280 =
=

Lowest Common Multiple (LCM)

LCM —
in question.

Find it from prime factors in two steps:

1. List
2.

EXAMPLE

Find the LCM of 8 and 14.

8 = ...
14 = ...

If a factor appears more than once in any number, list it that many times.

1.
2.

Highest Common Factor (HCF)

HCF —
in question.

Find it from prime factors in two steps:

1. List
2.

EXAMPLE

Find the HCF of 36 and 90.

36 = ...
90 = ...

1. 2.

You can also find LCM/HCF by listing the .... of both numbers and taking the ....

Fractions

First Go: ..... /..... /.....

Simplifying Fractions

To simplify, divide top and bottom by the ________ until they won't divide any more.

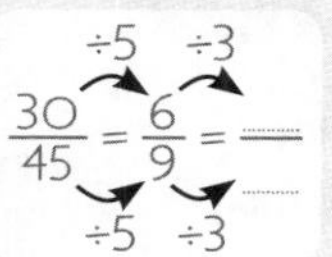

$\frac{30}{45} \overset{\div 5}{=} \frac{6}{9} \overset{\div 3}{=} \frac{\dots}{\dots}$

$\frac{42}{70} \overset{\div 7}{=} \frac{6}{10} \overset{\div 2}{=} \frac{\dots}{\dots}$

Mixed Numbers and Improper Fractions

MIXED NUMBER — has ________ part and ________ part, e.g. $2\frac{1}{3}$.

IMPROPER FRACTION — has ________ larger than ________, e.g. $\frac{7}{5}$.

To write mixed numbers as improper fractions:

1. Write as an ________.
2. Turn integer part into a ________.
3. ________ together.

$2\frac{3}{4} = 2 + \frac{3}{4} = \frac{\dots}{\dots} + \frac{3}{4} = \frac{\dots}{\dots}$
(1) (2) (3)

To write improper fractions as mixed numbers:

1. ________ top by bottom.
2. Answer is ________ part, ________ goes on top of fraction part.

(1) $17 \div 3 = 5$ remainder 2

(2) So $\frac{17}{3} = \dots\frac{\dots}{\dots}$

Multiplying and Dividing

1. Rewrite any mixed numbers as ________.
   If dividing → 2. Turn 2nd fraction ________. Change ÷ to ×.
   If multiplying → 3.
3. Cancel down with ________ factors.
4. ________ tops and bottoms separately.

EXAMPLE

Find $1\frac{3}{5} \div \frac{6}{7}$.

(1) $1\frac{3}{5} \div \frac{6}{7} = \frac{8}{5} \div \frac{6}{7}$

(2) $= \frac{\cancel{8}^{4}}{5} \times \frac{7}{\cancel{6}_{3}}$

(3) $= \frac{4}{5} \times \frac{7}{3}$

(4) = ________ = ________

Common Denominators

Use to ________, add or subtract fractions.

Find a number that all ________ divide into — the ________ is best.

EXAMPLE

Put $\frac{11}{6}, \frac{17}{12}, \frac{7}{4}$ in descending order.

LCM of 6, 12, 4 is 12.

$\frac{11}{6} = \frac{22}{12}$ $\frac{7}{4} = \frac{21}{12}$

$\frac{22}{12} > \frac{21}{12} > \frac{17}{12}$

So ____, ____, ____

Second Go: ...../...../.....

Fractions

Simplifying Fractions

To simplify, divide by the until they .

$\frac{30}{45} = \frac{\dots}{\dots} = \frac{\dots}{\dots}$ $\frac{42}{70} = \frac{\dots}{\dots} = \frac{\dots}{\dots}$

Mixed Numbers and Improper Fractions

MIXED NUMBER — has , e.g. $2\frac{1}{3}$.

IMPROPER FRACTION — has , e.g. $\frac{7}{5}$.

To write mixed numbers as improper fractions:

1 Write as

2 Turn

3

$2\frac{3}{4} =$ (1) = (2) = (3)

To write improper fractions as mixed numbers:

1

2 Answer is

(1) $17 \div 3 =$

(2) So $\frac{17}{3} =$

Multiplying and Dividing

1 Rewrite any

If multiplying

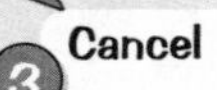
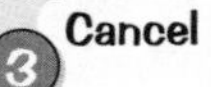
3 Cancel

If dividing

2 Turn

Change

4 Multiply

EXAMPLE

Find $1\frac{3}{5} \div \frac{6}{7}$.

(1) $1\frac{3}{5} \div \frac{6}{7} =$

(2) =

(3) =

(4) = =

Common Denominators

Use to , or fractions.

Find a number that all — the is best.

EXAMPLE

Put $\frac{11}{6}, \frac{17}{12}, \frac{7}{4}$ in descending order.

LCM of 6, 12, 4 is 12.

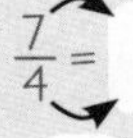
$\frac{11}{6} =$ $\frac{7}{4} =$

> >

So , ,

Fractions, Decimals and Percentages

First Go: ..... /..... /.....

Adding and Subtracting Fractions

1. Make the same.
2. Add/subtract the only.

EXAMPLE

Find $1\frac{1}{3} - \frac{5}{8}$.

$1\frac{1}{3} - \frac{5}{8} = \quad - \frac{5}{8} = \frac{32}{24} - \frac{15}{24}$ ①

Rewrite any mixed numbers.

② $= \frac{32-15}{24} =$

Fractions of Amounts

To find a fraction of a number:

1. Divide the number by the .
2. Multiply by the .

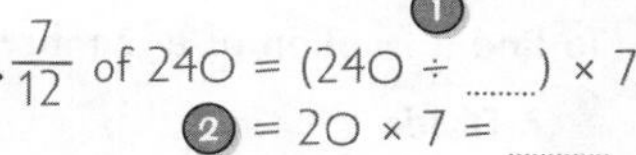

.... then .... if it's easier.

To write one number as a fraction of another:

1. Write the number over the .
2. Cancel down.

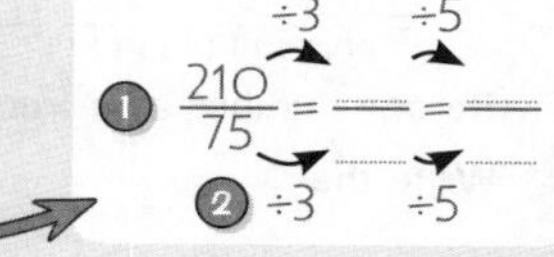

Common Conversions

Fractions, decimals and percentages are all ....
You can convert between them.

Fraction	Decimal	Percentage
	0.5	50%
$\frac{1}{4}$		25%
$\frac{3}{4}$	0.75	
$\frac{1}{3}$		$33\frac{1}{3}\%$
	0.6666...	$66\frac{2}{3}\%$

Fraction	Decimal	Percentage
$\frac{1}{10}$	0.1	
$\frac{1}{5}$		20%
$\frac{1}{8}$	0.125	
	0.375	37.5%
$\frac{5}{2}$		250%

How to Convert

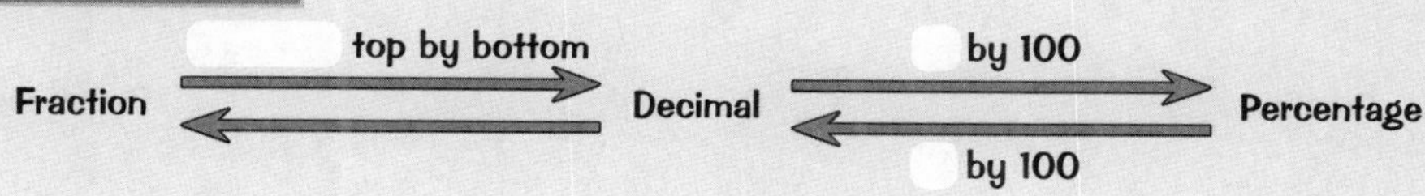

Second Go:
...../...../.....

Fractions, Decimals and Percentages

Adding and Subtracting Fractions

① Make

② Add/subtract the

EXAMPLE

Find $1\frac{1}{3} - \frac{5}{8}$.

$1\frac{1}{3} - \frac{5}{8} = \quad - \frac{5}{8} = \quad - \quad$ ①

Rewrite any mixed numbers.

② = =

Fractions of Amounts

To find a fraction of a number:

① Divide

② Multiply

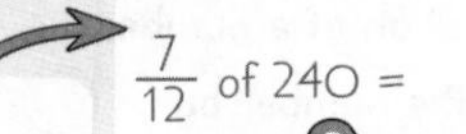

$\frac{7}{12}$ of 240 = ①

② =

.... if it's easier.

To write one number as a fraction of another:

① Write the

②

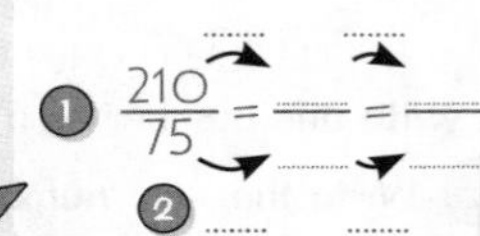

① $\frac{210}{75}$ = —— = ——

②

Common Conversions

Fractions, decimals and percentages are ....

You can ....

Fraction	Decimal	Percentage
	0.5	
$\frac{1}{4}$		
		75%
		$33\frac{1}{3}\%$
$\frac{2}{3}$		

Fraction	Decimal	Percentage
		10%
$\frac{1}{5}$		
	0.125	
$\frac{3}{8}$		
	2.5	

How to Convert

Fraction ⇄ Decimal ⇄ Percentage

Terminating and Recurring Decimals

First Go: ...../...../.....

Converting Terminating Decimals

TERMINATING DECIMALS — are (come to an end), e.g. 0.7 and 2.618.

When simplified, have only and as prime factors.

Three steps to write as fractions:

1. Put the digits after the decimal point as the .
2. Count the decimal places and put a with that many zeros as the .
3. down.

EXAMPLE

Write these decimals as fractions in their simplest forms:

a) 0.308

① $0.308 = \frac{308}{} = \frac{308}{1000}$ ②

③ $= \frac{}{500} =$

3 decimal places, so use $10^3 = 1000$.

b) 0.0015

① $0.0015 = \frac{15}{} = \frac{15}{10\,000}$ ②

③ $=$

4 decimal places, so use $10^4 = 10\,000$.

Converting Recurring Decimals

RECURRING DECIMALS — have a pattern of numbers that .

Repeated bit is marked with . One dot: that digit is repeated, e.g. $0.1\dot{6} = 0.166...$

Two dots: everything from the to is repeated, e.g. $0.\dot{1}8\dot{7} = 0.187187...$

To write a recurring decimal as a fraction:

1. Name the decimal r.
2. r by a power of 10 to get any parts past the decimal point.
3. by a power of 10 again to get one past the decimal point.
4. to get rid of the decimal part.
5. Divide and to find r.

EXAMPLE

Write $0.5\dot{7}$ as a fraction.

① Let $r = 0.5\dot{7}$

② $10r = 5.\dot{7}$

③ $100r =$ ....

$100r =$ ....

④ $-\ \ 10r = 5.\dot{7}$

$90r =$ ....

⑤ $r = \frac{......}{......} = \frac{......}{......}$

To write a fraction as a recurring decimal:

Find an with all in the denominator — the numerator is the part.

OR

Do the division (÷).

$\frac{14}{111} = \frac{126}{999} =$ .... (×9 top and bottom)

Second Go:
...../...../.....

Terminating and Recurring Decimals

Converting Terminating Decimals

TERMINATING DECIMALS — are

When simplified, denominators

Three steps to write as fractions:

1. Put
2. Count
3.

EXAMPLE

Write these decimals as fractions in their simplest forms:

a) 0.308

① 0.308 = ―― = ②

③ = $\frac{\quad}{500}$ =

b) 0.0015

① 0.0015 = ―― = ②

③ =

Converting Recurring Decimals

RECURRING DECIMALS — have
Repeated bit is
One dot:
Two dots:

To write a recurring decimal as a fraction:

1. Name
2. Multiply r by a power of 10 to get
3. Multiply by a power of 10 again to get
4. Subtract
5. to find r.

EXAMPLE

Write $0.5\dot{7}$ as a fraction.

① Let r = ...

② $10r$ = ...

③ $100r$ = ...

④ $100r$ = ...
$-\ \ 10r$ = ...
.... = ...

⑤ r = $\frac{\dots}{\dots}$ = $\frac{\dots}{\dots}$

To write a fraction as a recurring decimal:

Find an

— the

.

OR

Do the ( ).

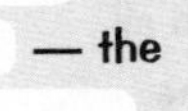

$\frac{14}{111}$ = $\frac{\dots}{\dots}$ = ...

Rounding and Estimating

First Go:
..... /..... /.....

Rounding to Decimal Places (d.p.) and Significant Figures (s.f.)

Digit after the last digit is the ____ :

- If decider is ____ , round last digit up.
- If decider is ____ , leave last digit as is.

To find significant figures:

1. The first ____ digit is the first s.f.
2. Each ____ ____ (zero or non-zero) is another s.f.

After rounding, fill in with ____ up to, not beyond, the ____ .

18.074 rounded to:

3 s.f.: Last digit = 0, so decider = ... — round up to ....

2 d.p.: Last digit = 7, so decider = ... — leave as ....

1 s.f.: Last digit = 1, so decider = ... — round up to .... Fill in with a 0.

Estimating Calculations and Square Roots

To estimate calculations, round all numbers to either ____ or ____ .

$$\frac{20.2 \times 2.87}{5.913} \approx \frac{\ldots \times \ldots}{\ldots} = \frac{60}{6} = \ldots$$

Two steps to estimate square roots:

1. Find a ____ on each side of the given number.
2. Decide which it's ____ , then ____ the digit after the decimal point.

EXAMPLE

Estimate the value of $\sqrt{68}$ to 1 d.p.

1. 64 (= 8^2) < 68 < .... (= ....^{2})
2. 68 is closer to 64 than ....., so $\sqrt{68}$ is closer to 8 than ..... $\sqrt{68} \approx$ ....

Upper and Lower Bounds

____ BOUND ≤ actual value < ____ BOUND

Rounded value − ____ unit ≤ actual value < Rounded value + ____ unit

Truncated value ≤ actual value < Truncated value + ____ unit

To find max and min values for a calculation:

1. Find ____ for each number.
2. Pick bounds to use for each ____ .

EXAMPLE

To 1 d.p., x = 1.4 and y = 3.7. What are the maximum and minimum values of $x \times y$?

1. ____ $\leq x <$ 1.45
   3.65 $\leq y <$ ____
   (1 d.p. is 0.1, so half of this is 0.05.)
2. max($x \times y$) = max(x) × max(y) = 1.45 × ____ = 5.4375
   min($x \times y$) = min(x) × min(y) = ____ × 3.65 = 4.9275

max(a ÷ b) = max(a) ÷ ....

min(a ÷ b) = ... ÷ max(b)

Second Go: ...../...../.....

Rounding and Estimating

Rounding to Decimal Places (d.p.) and Significant Figures (s.f.)

is the decider:

- If decider is .
- If decider is .

To find significant figures:

1. is the first s.f.
2. Each is another s.f.

After rounding, fill in

18.074 rounded to:

3 s.f.: Last digit = ..., so decider = ...
— ....

2 d.p.: Last digit = ..., so decider = ...
— ....

1 s.f.: Last digit = ..., so decider = ...
— ....

Estimating Calculations and Square Roots

To estimate calculations,

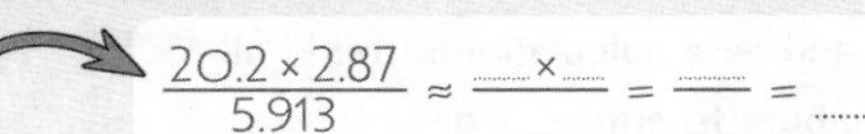

$\frac{20.2 \times 2.87}{5.913} \approx \frac{\dots \times \dots}{\dots} = \frac{\dots}{\dots} = \dots$

Two steps to estimate square roots:

1. Find
2. Decide

EXAMPLE

Estimate the value of $\sqrt{68}$ to 1 d.p.

1. .... < 68 < ....
2. 68 is closer to ...., so $\sqrt{68}$ is closer to .... $\sqrt{68} \approx$ ....

Upper and Lower Bounds

BOUND ≤ actual value < BOUND

Rounded value ≤ actual value <

≤ actual value < Truncated value

To find max and min values for a calculation:

1. Find
2. Pick

EXAMPLE

To 1 d.p., $x = 1.4$ and $y = 3.7$. What are the maximum and minimum values of $x \times y$?

1. $\le x <$
   $\le y <$

   1 d.p. is 0.1, so half of this is 0.05.

2. $\max(x \times y) =$
   $=$
   $\min(x \times y) =$
   $=$

max(a ÷ b) = ...

min(a ÷ b) = ...

 ☐  ☐  ☐

Standard Form

First Go: ..... /..... /.....

Numbers in Standard Form

Number between and → $A \times 10^n$ ← Number of places the moves — for big numbers, for small numbers

EXAMPLE

What is 70.6 million in standard form?

70.6 million = 70.6 × ....

= 70 600 000.0 — Count how far the decimal point moves.

= 7.06 × 10.... — Big number, so positive

EXAMPLE

Express 5.129×10^{-4} as an ordinary number.

Negative, so small number

00005.129×10^{-4} — Move the decimal point by this many places.

= ....

Three Steps to Multiply or Divide

1. so the front numbers and powers of 10 are together.
2. Multiply/divide the front numbers. Use to multiply/divide the powers of 10.
3. Put the answer in .

EXAMPLE

Find $(8.15 \times 10^7) \times (4 \times 10^{-3})$. Give your answer in standard form.

$(8.15 \times 10^7) \times (4 \times 10^{-3})$

1. $= (\quad \times \quad) \times (10^7 \times 10^{-3})$
2. $= 32.6 \times 10^{7+-3}$ — Add the powers.
   =
3. $= 3.26 \times 10 \times 10^4$
   =

Three Steps to Add or Subtract

1. Rewrite so the are the same.
2. Add/subtract .
3. Put the answer in .

EXAMPLE

Find $(3.4 \times 10^{-6}) + (9.7 \times 10^{-5})$. Give your answer in standard form.

$(3.4 \times 10^{-6}) + (9.7 \times 10^{-5})$

1. $= (\quad \times 10 \times 10^{-6}) + (9.7 \times 10^{-5})$
   $= (\quad \times 10^{-5}) + (9.7 \times 10^{-5})$
2. $= (\quad + \quad) \times 10^{-5}$
   $= \quad \times 10^{-5}$ — Not in standard form yet — front number is bigger than 10.
3. $= \quad \times 10 \times 10^{-5}$
   =

Second Go:
...../...../.....

Standard Form

Numbers in Standard Form

Number

Number

— positive for

EXAMPLE

What is 70.6 million in standard form?

70.6 million = ...

= 70 600 000.0

= ....

EXAMPLE

Express 5.129×10^{-4} as an ordinary number.

00005.129×10^{-4}

= ....

Three Steps to Multiply or Divide

1. Rearrange so
2. Multiply/divide
   Use
3. Put

EXAMPLE

Find $(8.15 \times 10^7) \times (4 \times 10^{-3})$.
Give your answer in standard form.

$(8.15 \times 10^7) \times (4 \times 10^{-3})$

1. = ×
2. = × 10
   = × 10
3. =
   =

Three Steps to Add or Subtract

1. Rewrite so
2. Add/subtract
3. Put

EXAMPLE

Find $(3.4 \times 10^{-6}) + (9.7 \times 10^{-5})$.
Give your answer in standard form.

$(3.4 \times 10^{-6}) + (9.7 \times 10^{-5})$

1. = $+ (9.7 \times 10^{-5})$
   = $+ (9.7 \times 10^{-5})$
2. = × 10
   = × 10
3. =
   =

Mixed Practice Quizzes

100% of these four quizzes cover p.3-16. That's... give me a sec... sorry, this is so embarrassing... four quizzes. Whew. Mark each one and tot up your score.

Quiz 1

Date: ...... / ...... / ......

1) Which bound could be the actual value of a rounded number: upper or lower?
2) How do you find the HCF of two numbers using their prime factors?
3) What would be the first step in writing a terminating decimal as a fraction?
4) What do you do when dividing fractions that you wouldn't do when multiplying them?
5) How do you find a fraction of a number?
6) Which factors of 32 are missing from this list: 2, 8, 32?
7) What do the letters BODMAS stand for?
8) True or false? After rounding, fill in with zeros only up to the decimal point.
9) Is 0.94×10^8 in standard form?
10) What is a prime number?

Total:

Quiz 2

Date: ...... / ...... / ......

1) What is the integer part when $\frac{31}{4}$ is converted into a mixed number?
2) If a number is in standard form, between which two values should its first number be?
3) Is the product of two negative numbers positive or negative?
4) What type of number can't be written as a fraction?
5) What should you circle in a factor tree?
6) How do you write 0.126126126... using recurring decimal notation?
7) What do you need to do to write 160 as a fraction of 85?
8) What is the first step in finding the answer to $1.7 \times 10^3 + 4.8 \times 10^5$?
9) What is usually the best choice for a common denominator of two or more fractions?
10) How do you estimate the answer to a calculation?

Total:

Mixed Practice Quizzes

Quiz 3

Date: ...... / ...... / ......

1) How do you write a percentage as a decimal?
2) When rounded to 1 d.p., a number is 12.4. What is the lower bound of its actual value?
3) When should you stop listing pairs of numbers when finding factors?
4) What would be the first step in multiplying two numbers in standard form?
5) What equivalent fraction should you find to write a fraction as a recurring decimal?
6) What is the first step in estimating the value of a square root?
7) What is a lowest common multiple?
8) In the calculation $4 \times 3 + 2$, what should be carried out first: the addition or the multiplication?
9) What is the first thing you need to do to add $\frac{3}{8}$ and $\frac{1}{6}$?
10) How do you convert a mixed number into an improper fraction?

Total:

Quiz 4

Date: ...... / ...... / ......

1) Is the answer to $-12 \div 6$ positive or negative?
2) What two bounds should you use to find the minimum value of $x \div y$?
3) True or false? 0.0025 in standard form is 2.5×10^3.
4) When a terminating decimal is written as a simplified fraction, what are the only possible prime factors of the denominator?
5) What is 0.375 as: a) a percentage? b) a fraction?
6) In a factor tree for 450, the circled numbers are 5, 3, 5, 2 and 3. Write the prime factorisation of 450 using index notation.
7) What is the first step in multiplying two mixed numbers?
8) What is 10.7548 rounded to 3 s.f.?
9) How do you simplify a fraction?
10) What is the next step in writing a recurring decimal as a fraction after calling the decimal r?

Total:

Algebra Basics

First Go: ..... /..... /.....

Algebraic Notation

Only q is .... — not p.

Notation	Meaning
abc	a b c
$\frac{a}{b}$	a b
pq^3	p × × ×
$(mn)^2$	m × m × ×
x( )^3	x × (y – z) × (y – z) × (y – z)

Brackets mean both m and n are ....

Things like -4^2 are unclear. Write either (.....)^2 = 16 or –(.....) = –16 instead.

Collecting Like Terms

TERM — a collection of numbers, and , all multiplied/divided together.

1. Put around each term.
2. Move so like terms are .
3. like terms.

EXAMPLE

Simplify $7a + 2 - 3a + 5$.

1. $7a$ $+2$ $-3a$ $+5$ — Put the +/– sign in each bubble.
2. $= 7a$ $-3a$ $+2$ $+5$
3. =

Ten Rules for Powers

These are only true for powers of the ....

1. Multiplying — powers: e.g. $a^2 \times a^5 =$
2. Dividing — powers: e.g. $b^5 \div b^3 =$
3. Raising one power to another — powers: e.g. $(p^2)^4 =$
4. Anything to the power of 1 is : e.g. $x^1 =$
5. Anything to the power of 0 is : e.g. $y^0 =$
6. 1 to the power of anything is still : e.g. $1^x =$
7. Apply powers to the TOP and BOTTOM of e.g. $\left(\frac{m}{n}\right)^2 =$
8. NEGATIVE powers — , then make the power . $5^{-2} = \frac{1}{5^2} = \frac{.....}{.....}$
9. FRACTIONAL powers are — e.g. power of $\frac{1}{2}$ is a square root, power of $\frac{1}{3}$ is a cube root, etc. $16^{\frac{1}{4}} = \sqrt[4]{16} =$ .....
10. TWO-STAGE FRACTIONAL powers — do the root, then the . $27^{\frac{2}{3}} = (27^{.....})^2 = 3^2 =$ .....

Second Go: ...../...../.....

Algebra Basics

Algebraic Notation

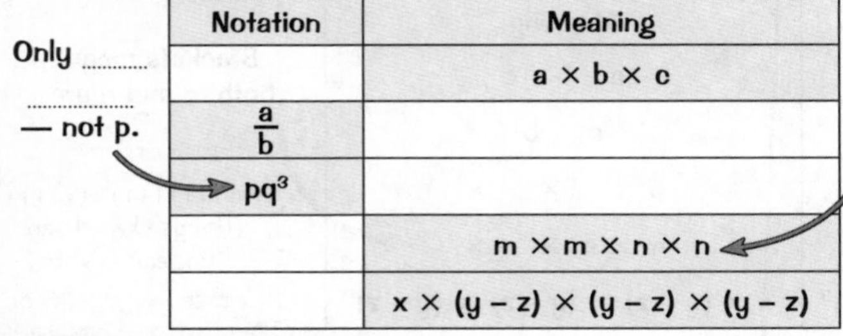

Only .... — not p.

Notation	Meaning
	a × b × c
$\frac{a}{b}$	
pq³	
	m × m × n × n
	x × (y – z) × (y – z) × (y – z)

Brackets mean ....

Things like -4^2 are unclear. Write either .... or .... instead.

Collecting Like Terms

TERM — a collection of

1. Put
2. Move
3. Combine

EXAMPLE

Simplify $7a + 2 - 3a + 5$.

Put the +/– sign in each bubble.

1.
2. =
3. =

Ten Rules for Powers

These are only true for ....

1. Multiplying — : e.g. $a^2 \times a^5 =$
2. Dividing — : e.g. $b^5 \div b^3 =$
3. one power to another — : e.g. $(p^2)^4 =$
4. Anything to the power of is : e.g. $= x$
5. to the power of 0 is : e.g. $y^0 =$
6. 1 to the power of is still : e.g. $= 1$
7. Apply powers of fractions e.g. $\left(\frac{m}{n}\right)^2 =$
8. NEGATIVE powers — , then . $5^{-2} =$
9. FRACTIONAL powers are — e.g. power of $\frac{1}{2}$ is a , power of $\frac{1}{3}$ is a etc.  $16^{\frac{1}{4}} =$
10. TWO-STAGE FRACTIONAL powers — , then . $27^{\frac{2}{3}} =$

Expanding Brackets and Factorising

First Go: ..... /..... /.....

Expanding Brackets

Multiply everything [blank] the bracket by everything [blank] the bracket.

$2x(5 - 3y) = (2x \times 5) + (2x \times -3y)$
$= \text{.................}$

The FOIL method for [blank]:

- Multiply .... terms of each bracket.
- Multiply .... terms together.
- Multiply .... terms together.
- Multiply .... terms of each bracket.

$(m - 6)(3m + 4)$ (F, O, I, L)

$= (m \times 3m) + (\text{..........})$
$+ (-6 \times 3m) + (\text{..........})$
$= 3m^2 + \text{.....} - 18m - \text{.....}$
$= \text{.....................}$

To multiply out triple brackets, ... two together as normal, then multiply the result by the .... bracket.

Factorising Expressions

FACTORISING — putting [blank] back in.

1. Take out [blank] that goes into all terms.
2. Take out [blank] of each letter that goes into all terms.
3. Open bracket and [blank] what's needed to reproduce the [blank].
4. [blank] your answer.

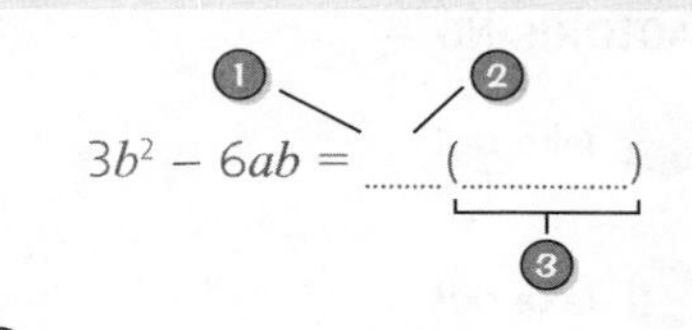

$3b^2 - 6ab = \text{.....}(\text{..........})$

4 .... $= 3b \times b - 3b \times 2a$
$= 3b^2 - 6ab$

The bits put in front of the bracket are the ....

The Difference of Two Squares (D.O.T.S.)

D.O.T.S. — 'one thing squared' take away '[blank]'.

Use this rule for factorising: $a^2 - b^2 = (\text{..........})(\text{..........})$

EXAMPLE

Factorise $5p^2 - 20q^2$.

$5p^2 - 20q^2 = 5(p^2 - 4q^2) = \text{.....................}$

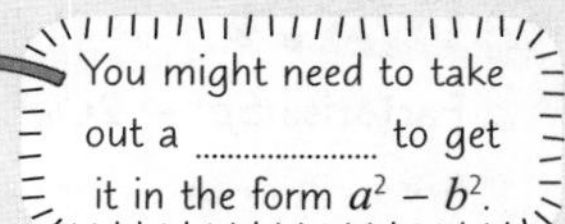

You might need to take out a .... to get it in the form $a^2 - b^2$.

Second Go: ...../...../.....

Expanding Brackets and Factorising

Expanding Brackets

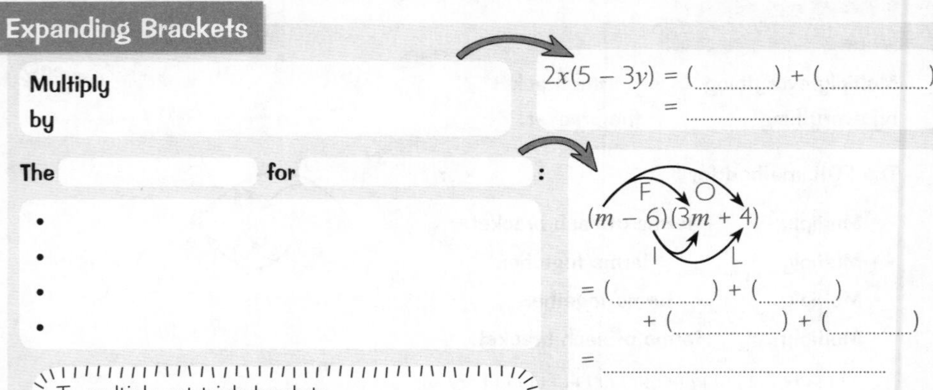

Factorising Expressions

FACTORISING — .

1. Take out
2. Take out
3. Open bracket and
4. Check

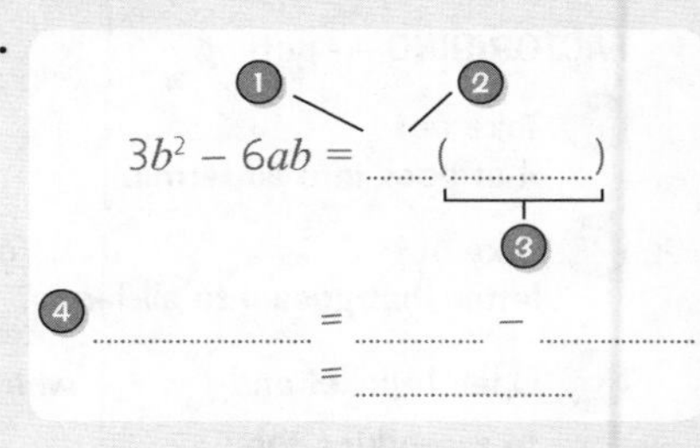

The bits put in front of the ... are the ... .

The Difference of Two Squares (D.O.T.S.)

D.O.T.S. — .

Use this rule for factorising:

EXAMPLE

Factorise $5p^2 - 20q^2$.

$5p^2 - 20q^2 = \dots = \dots$

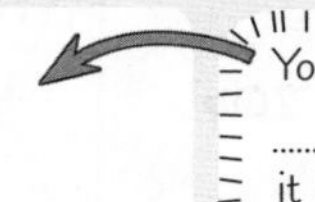

You might need to ... ... to get it in the form ... .

Surds and Solving Equations

First Go: ..... /..... /.....

Six Rules for Manipulating Surds

1. $\sqrt{a} \times \sqrt{b} = \sqrt{a \times b}$

   $\sqrt{5} \times \sqrt{3} =$ ....

2. $\frac{\sqrt{a}}{\sqrt{b}} =$

   $\frac{\sqrt{27}}{\sqrt{3}} = \sqrt{\frac{27}{3}} = \sqrt{9} =$ ....

3. $\sqrt{a} + \sqrt{b}$ — .
   (Definitely NOT $\sqrt{a+b}$)

4. $(a + \sqrt{b})^2 =$

   $(6 + \sqrt{2})^2 = (6 + \sqrt{2})(6 + \sqrt{2})$
   $= 36 + 12\sqrt{2} + 2$
   $=$ ....

5. $(a + \sqrt{b})(a - \sqrt{b}) =$

   $(4 + \sqrt{7})(4 - \sqrt{7}) = 16 + 4\sqrt{7} - 4\sqrt{7} - 7$
   $=$ .... $=$ ....

6. $\frac{a}{\sqrt{b}} =$

   $\frac{3}{\sqrt{5}} = \frac{3}{\sqrt{5}} \times \frac{\ldots}{\ldots} = \frac{\ldots}{\ldots}$

This is known as '.. the denominator'.

EXAMPLE

Write $\sqrt{54} + \sqrt{150} - \sqrt{24}$ in the form $a\sqrt{6}$.

$\sqrt{54} = \sqrt{9 \times 6} = \sqrt{9} \times \sqrt{6} =$ ....

$\sqrt{150} = \sqrt{25 \times 6} = \sqrt{25} \times \sqrt{6} =$ ....

$\sqrt{24} = \sqrt{4 \times 6} = \sqrt{4} \times \sqrt{6} =$ ....

.... + ... − .... = ...

Six Steps to Solve Equations

1. **Get rid of .**
2. **Multiply out .**
3. **Put on one side, on the other.**
4. **Reduce to the form .**
5. **Divide by A to get ' '.**
6. **If you have ' ' instead, both sides.**

You can ignore .... that don't apply to the equation.

EXAMPLE

Solve $\frac{3}{x-2} = \frac{2}{3x+1}$.

1.
2. $9x + 3 = 2x - 4$
3. $9x - 2x = -4 - 3$
4. $7x = -7$
5.

There's no x^2, so stop at Step 5.

EXAMPLE

Solve $x(5x) - 13 = 7$.

2. $5x^2 - 13 = 7$
3. $5x^2 = 7 + 13$
4.
5. $x^2 = 4$
6.

Taking the square root gives a .... and a .... solution.

Second Go: ...../...../.....

Surds and Solving Equations

Six Rules for Manipulating Surds

1. $\sqrt{a} \times \sqrt{b} =$

$\sqrt{5} \times \sqrt{3} =$ ....

2. $\dfrac{\sqrt{a}}{\sqrt{b}} =$

$\dfrac{\sqrt{27}}{\sqrt{3}} = \sqrt{\dfrac{\dots}{\dots}} = \dots = \dots$

3. .... — ....

(Definitely NOT $\sqrt{a+b}$)

4. $(a+\sqrt{b})^2 =$

$(6+\sqrt{2})^2 = ($..........$)($..........$)$

= ....

= ....

5. $(a+\sqrt{b})(a-\sqrt{b}) =$

$(4+\sqrt{7})(4-\sqrt{7}) =$ ....

= .... = ...

6. $\dfrac{a}{\sqrt{b}} =$

$\dfrac{3}{\sqrt{5}} = \dfrac{\dots}{\dots} \times \dfrac{\dots}{\dots} = \dfrac{\dots}{\dots}$

This is known as '..........

..........'.

EXAMPLE

Write $\sqrt{54} + \sqrt{150} - \sqrt{24}$ in the form $a\sqrt{6}$.

$\sqrt{54} =$

$\sqrt{150} =$

$\sqrt{24} =$

.... = ...

Six Steps to Solve Equations

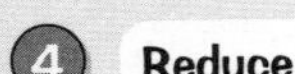

1.

2.

3. Put

4. Reduce

5. Divide

6. If you have '$x^2 = \dots$' instead,

You can ignore ....

....

EXAMPLE

Solve $\dfrac{3}{x-2} = \dfrac{2}{3x+1}$.

1.
2.
3.
4.
5. $x =$

There's no x^2, so stop at Step 5.

EXAMPLE

Solve $x(5x) - 13 = 7$.

2.
3.
4.
5. $x^2 =$
6. $x =$

Taking the square root gives ......

....

....

....

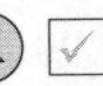

Rearranging Formulas

First Go: ...../...../.....

Seven Steps for Rearranging Formulas

1. Get rid of ______.
2. Get rid of ______.
3. Multiply out ______.
4. Put ______ on one side, ______ on the other.
5. Reduce to the form ______ (where x is the ______).
6. Divide by A to get '______'.
7. If you're left with '______', ______ both sides.

A and B could be numbers, ... or ....

If the Subject is in a Fraction

EXAMPLE

Make p the subject of $q = \frac{7p - 3}{5}$.

2. ______
4. $7p = 5q + 3$
5. This is in the form $Ap = B$.
6. $p =$ ______

No square roots or brackets, so ignore Steps 1 and 3.

If there's a Square or Square Root

EXAMPLE

Make r the subject of $s^2 = 9 - 3r^2$.

4. ______
5. This is in the form $Ar^2 = B$.
6. $r^2 = \frac{9 - s^2}{3}$
7. $r =$ ______

No square roots, fractions or brackets, so ignore Steps 1-3.

If the Subject Appears Twice

You'll need to factorise, usually at Step 5.

EXAMPLE

Make m the subject of $n = \frac{m}{m - 3}$.

2. ______
3. $mn - 3n = m$
4. $mn - m = 3n$
5. ______

This is where you factorise — m is a common factor.

6. $m =$ 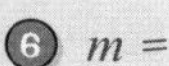

No square roots, so ignore Step 1.

EXAMPLE

Make a the subject of $2b + 1 = \sqrt{4a - 3}$.

1. ______
3. $4b^2 + 4b + 1 = 4a - 3$
4. $4a = 4b^2 + 4b + 4$
5. This is in the form $Aa = B$.
6. $a =$ ______

No fractions, so ignore Step 2.

Second Go: ...../...../.....

Rearranging Formulas

Seven Steps for Rearranging Formulas

1.
2.
3.
4. Put
5. Reduce (where).
6. Divide
7. If you're left with

A and B could be ....

....

If the Subject is in a Fraction

EXAMPLE

Make p the subject of $q = \frac{7p - 3}{5}$.

2.
4.
6. $p =$

5. This is in the form $Ap = B$.

No square roots or brackets, so ignore Steps 1 and 3.

If there's a Square or Square Root

EXAMPLE

Make r the subject of $s^2 = 9 - 3r^2$.

4.
6.
7. $r =$

5. This is in the form $Ar^2 = B$.

No square roots, fractions or brackets, so ignore Steps 1-3.

EXAMPLE

Make a the subject of $2b + 1 = \sqrt{4a - 3}$.

1.
3.
4.
6. $a =$

5. This is in the form $Aa = B$.

No fractions, so ignore Step 2.

If the Subject Appears Twice

You'll need to ...., usually at Step 5.

EXAMPLE

Make m the subject of $n = \frac{m}{m - 3}$.

2.
3.
4.
5.
6. $m =$

This is where you factorise — m is a common factor.

No square roots, so ignore Step 1.

Mixed Practice Quizzes

Expand your algebra knowledge with these quick quizzes and see what you've learned from p.19-26. Once you've checked your answers, jot down your score.

Quiz 1

Date: ...... / ...... / ......

1) How do you expand triple brackets?
2) What is $a \times a \times b$ in algebraic notation?
3) True or false? $(x + \sqrt{y})^2 = x^2 + y$.
4) Collect like terms to simplify the expression $3x - 7y + 4x - 2y$.
5) What term goes outside the bracket when $6x^2 + 27xy^2$ is fully factorised?
6) How do you raise a fraction to a power?
7) If $\sqrt{75}$ is written in the form $a\sqrt{3}$, what is the value of a?
8) What is the first step you'd take to solve an equation involving a fraction?
9) What is the first step needed to make y the subject of $x = 3 - y^2$?
10) What is the value of $8^{-\frac{1}{3}}$?

Total:

Quiz 2

Date: ...... / ...... / ......

1) How do you multiply two surds together?
2) Simplify the expression $4a \times 3 \times 2a$.
3) How do you divide two powers of the same number?
4) What is $(2 + \sqrt{5})(2 - \sqrt{5})$ in its simplest form?
5) How do you use the FOIL method to expand double brackets?
6) What is the first thing you'd do when rearranging $2q - 1 = \sqrt{p - 5}$ to make p the subject?
7) What is the value of 1^5?
8) What goes outside the brackets when factorising an expression?
9) Solve the equation $2x^2 = 18$.
10) How do you raise a number to a negative power?

Total:

Mixed Practice Quizzes

Quiz 3

Date: ...... / ...... / ......

1) What is the definition of a term?
2) Solve $4x - 3 = 6x + 5$.
3) How do you expand a single term over a bracket?
4) Factorise $16 - y^2$.
5) How do you raise a power to another power?
6) What is the first step you'd do to make b the subject of $a = \frac{3 - b}{5}$?
7) Give the rule for dividing surds.
8) What is x^0?
9) What is meant by 'rationalising the denominator'?
10) Which operation is the same as raising a number to the power of $\frac{1}{2}$?

Total:

Quiz 4

Date: ...... / ...... / ......

1) What would the x^3 term be when $2x^2(5y^2 - 4x)$ is expanded?
2) How do you multiply two powers of the same number?
3) What is meant by 'the difference of two squares'?
4) What is the first step needed to solve $5(2x + 1) = 3(2x - 1)$?
5) If the subject appears in a formula twice, what extra step will you need to do when rearranging?
6) How is xy^3 different to $(xy)^3$?
7) How would you rationalise the denominator of $\frac{p}{\sqrt{q}}$?
8) What happens when you raise a number to the power of 1?
9) Rearrange $y = 4x - 11$ to make x the subject.
10) What is $125^{\frac{2}{3}}$?

Total:

Factorising Quadratics

First Go: ...../...../.....

Quadratic Equations

Standard form of a quadratic equation: .... + ... + ... = 0

a, b and c can be any ....

To FACTORISE — put it into .

To SOLVE — find the that make each bracket .

Factorising when a = 1

1. Rearrange to .
2. Write two brackets:
3. Find two numbers that to give 'c' AND to give 'b'.
4. Fill in signs.
5. Check by .
6. Solve the equation.

EXAMPLE

Solve $x^2 - 6x = -8$.

1. $x^2 - 6x + 8 = 0$
2. $(x \quad)(x \quad) = 0$
3. Factor pairs of 8: or
   $2 + 4 = 6$, so you need 2 and 4.
4. $(x - 2)(x - 4) = 0$
5. $(x - 2)(x - 4) = x^2 - 4x - 2x + 8$
   $=$
6. $(x - 2) = 0 \Rightarrow$
   $(x - 4) = 0 \Rightarrow$

Factorising when a is not 1

1. Rearrange to .
2. Write where the first terms to give 'a'.
3. Find pairs of numbers that to give 'c'.
4. Test each pair in to find one that adds/subtracts to give .
5. Fill in signs.
6. Check by .
7. Solve the equation.

EXAMPLE

Solve $2x^2 + x - 6 = 0$.

1. This is in the standard format.
2. $(\quad)(\quad) = 0$
3. Factor pairs of 6: or
4. $(2x \quad 1)(x \quad 6) \rightarrow 12x$ and x
   $(2x \quad 6)(x \quad 1) \rightarrow 2x$ and $6x$
   $(2x \quad 2)(x \quad 3) \rightarrow 6x$ and $2x$
   $(2x \quad 3)(x \quad 2) \rightarrow 4x$ and $3x$ — $4x - 3x = x$
5. $(2x - 3)(x + 2) = 0$
6. $(2x - 3)(x + 2)$
   $= 2x^2 + 4x - 3x - 6$
   $=$
7. $(2x - 3) = 0 \Rightarrow$
   $(x + 2) = 0 \Rightarrow$

Second Go: ..... /..... /.....

Factorising Quadratics

Quadratic Equations

Standard form of a quadratic equation:

a, b and c can be ....

To FACTORISE — .

To SOLVE — .

Factorising when a = 1

1. Rearrange
2. Write
3. Find two numbers
4. Fill in
5. Check
6. Solve the equation.

EXAMPLE

Solve $x^2 - 6x = -8$.

1
2
3 Factor pairs of 8: ...., so you need ....
4 (   )(   ) = 0
5 (   )(   )
=
=
6

Factorising when a is not 1

1. Rearrange
2. Write
3. Find pairs of numbers
4. Test each pair
5. Fill in
6. Check
7. Solve the equation.

EXAMPLE

Solve $2x^2 + x - 6 = 0$. 1 This is in the standard format.

2
3 Factor pairs of 6: ....
4 $(2x \quad 1)(x \quad 6)$ → ....
$(2x \quad 6)(x \quad 1)$ → ....
$(2x \quad 2)(x \quad 3)$ → ....
$(2x \quad 3)(x \quad 2)$ → ....
$4x - 3x = x$
5 (   )(   ) = 0
6 (   )(   )
=
=
7

Solving Quadratics

First Go: ...../...../.....

The Quadratic Formula

$$x = \frac{\quad \pm \sqrt{\quad}}{\quad}$$

Use the quadratic formula when:
- the quadratic .
- the question mentions or
- you need or .

1. Rearrange equation into the form .
2. Identify .
3. into formula.
4. Evaluate both .

Check your answers by substituting back into the ....

EXAMPLE

Find the solutions to $4x^2 + 3x = 5$ to 2 d.p.

1. $4x^2 + 3x - 5 = 0$
2. a = ..., b = ..., c = ...
3. $x = \frac{-3 \pm \sqrt{3^2 - 4 \times 4 \times -5}}{2 \times 4} = \frac{-3 \pm \sqrt{89}}{8}$
4. x = ... (2 d.p.) or .... (2 d.p.)

The ± sign means you get two solutions.

Completing the Square

1. Multiply out initial bracket
2. adjusting number to match original equation.
3. Set and solve.

EXAMPLE

Solve $x^2 + 4x - 3 = 0$.

Check this is in the standard format first.

1. $(x + 2)^2 =$
2. $(x + 2)^2 - 7 = x^2 + 4x + 4 - 7$
   $= x^2 + 4x - 3$
3. $(x + 2)^2 - 7 = 0$
   $(x + 2)^2 = 7$
   $x + 2 = \pm\sqrt{7}$, so $x =$

Add/subtract to get –3.

... when a is not 1

1. Take out a from the first two terms.
2. Multiply out initial bracket
3. adjusting number to match original equation.

EXAMPLE

Write $2x^2 - 8x + 3$ in the form $a(x + m)^2 + n$.

Check this is in the standard format first.

1.
2. $2(x - 2)^2 = 2x^2 - 8x + 8$
3. $2(x - 2)^2 - 5 = 2x^2 - 8x + 8 - 5$
   $= 2x^2 - 8x + 3$

Add/subtract to make this 3.

When a is positive, the adjusting number tells you the .... y-value of the graph. This occurs when the ...., i.e. when $x = -m$. This also gives you the coordinates of the .... of the graph.

Second Go:
...../...../.....

Solving Quadratics

The Quadratic Formula

$x =$

Use the quadratic formula when:
-
-
-

1. Rearrange
2.
3.
4. Evaluate

Check your answers by .... into the ....

EXAMPLE

Find the solutions to $4x^2 + 3x = 5$ to 2 d.p.

1. .... $= 0$
2. ...., ..., ...
3. $x =$ $=$
4. $x =$ .... (2 d.p.) or .... (2 d.p.)

The ± sign means you get two solutions.

Completing the Square

1. Multiply
2. Add/subtract
3.

EXAMPLE

Solve $x^2 + 4x - 3 = 0$.

Check this is in the standard format first.

1. $(\quad)^2 =$
2. $=$
   $=$
3. $= 0$
   , so $x =$

Add/subtract to get −3.

... when a is not 1

1. Take out a factor
2. Multiply
3. Add/subtract

EXAMPLE

Write $2x^2 - 8x + 3$ in the form $a(x + m)^2 + n$.

Check this is in the standard format first.

1. $2(\quad) + 3$
2.
3. $=$
   $=$

Add/subtract to make this 3.

When a is positive, the adjusting number tells you .... This occurs when ...., i.e. when $x =$ .... This also gives you the .... of the graph.

Algebraic Fractions

First Go: ..... /..... /.....

Simplifying Algebraic Fractions

terms on the top and bottom.
Deal with one or at a time.

$$\frac{8x^3y}{2x^2y^3} = \frac{{}^4\cancel{8} \times \cancel{x^3}^{x} \times \cancel{y}}{\cancel{2} \times \cancel{x^2} \times \cancel{y^3}_{y^2}}$$

÷2 on top and bottom
$\div x^2$ on top and bottom
$\div y$ on top and bottom

$= \frac{\dots\dots}{\dots\dots}$

You might have to first, then cancel a common factor:

$$\frac{x^2 - x - 2}{x^2 + 5x + 4} = \frac{\cancel{(x+1)}(x-2)}{\cancel{(x+1)}(x+4)}$$

$= \frac{\dots\dots\dots\dots}{\dots\dots\dots\dots}$

Multiplying

Multiply and of the fractions .

$$\frac{x}{3x+12} \times \frac{2x+8}{x-1}$$

$$= \frac{x}{3\,\cancel{(x+4)}} \times \frac{2\,\cancel{(x+4)}}{x-1}$$

$$= \frac{x \times 2}{3 \times (x-1)}$$

$=$

.... and cancel first, to make multiplying easier.

Dividing

To divide, turn the second fraction , then .

$$\frac{x-5}{x^2-9} \div \frac{5x}{x-3} = \frac{x-5}{x^2-9} \times$$

$$= \frac{x-5}{\cancel{(x-3)}(x+3)} \times \frac{\cancel{x-3}}{5x}$$

Factorise using D.O.T.S.

$$= \frac{x-5}{(x+3) \times 5x}$$

$=$

Adding and Subtracting

1. Find a .
2. Multiply the and of each fraction by whatever gives the .
3. Add or subtract .

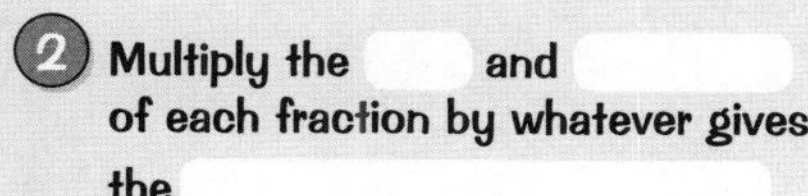
The common denominator is something .... divide into.

EXAMPLE

Write $\frac{2}{2x-1} - \frac{3}{x+4}$ as a single fraction in its simplest form.

1. Common denominator:
2. $\frac{2(x+4)}{(2x-1)(x+4)} - \frac{3(2x-1)}{(x+4)(2x-1)}$
3. $= \frac{2(x+4) - 3(2x-1)}{(2x-1)(x+4)} = \frac{2x+8-6x+3}{(2x-1)(x+4)}$

$=$ Collect like terms together.

Second Go:
...../...../.....

Algebraic Fractions

Simplifying Algebraic Fractions

Cancel terms
Deal with

$$\frac{8x^3y}{2x^2y^3} = \frac{{}^4\cancel{8} \times \cancel{x^3}^{\,x} \times \cancel{y}}{\cancel{2} \times \cancel{x^2} \times \cancel{y^3}_{\,y^2}}$$

.... on top and bottom
.... on top and bottom
.... on top and bottom

$= \frac{\dots\dots}{\dots\dots}$

You might have to

$$\frac{x^2 - x - 2}{x^2 + 5x + 4} = \frac{\cancel{(x+1)}(\dots\dots)}{\cancel{(x+1)}(\dots\dots)}$$

$= \frac{\dots\dots}{\dots\dots}$

Multiplying

Multiply

$$\frac{x}{3x+12} \times \frac{2x+8}{x-1}$$

$= \quad \times$

$=$

$=$

....
................, to make multiplying easier.

Dividing

To divide,

$$\frac{x-5}{x^2-9} \div \frac{5x}{x-3} = \quad \times$$

Factorise using D.O.T.S. $= \quad \times$

$=$

$=$

Adding and Subtracting

1

2 **Multiply the top and bottom**

3

The common denominator is ....
....

EXAMPLE

Write $\frac{2}{2x-1} - \frac{3}{x+4}$ as a single fraction in its simplest form.

1

2 $-$

3 $=$

$= \quad =$

Collect like terms together.

Sequences

First Go:
..... /..... /.....

nth term of Linear Sequences

LINEAR SEQUENCES — increase/decrease by ______ each time (common difference).

1. Find the ______ — this is what you multiply n by.
2. Work out what to ______.
3. Put ______ together.

EXAMPLE

Find the nth term of the sequence 7, 11, 15, 19 ...

1. 11 – 7 = __, 15 – 11 = __, etc.
   So common difference = __
2. For n = 1, 4n = 4. 7 – 4 = __,
   so __ is added to each term.
3. So nth term is ______

nth term of Quadratic Sequences

QUADRATIC SEQUENCES — have an __ term, so the difference between terms ______.

1. Find ______ between pairs of terms.
2. Find difference between ______.
3. ______ to get coefficient of n^2.
4. Subtract n^2 term (including ______) from each term to get a ______.
5. Find nth rule of the ______.
6. Put n^2 term and ______ together.

EXAMPLE

Find the nth term of the sequence 10, 14, 22, 34 ...

10 14 22 34

1. __ __ __
2. +4 +4
3. 4 ÷ 2 = 2,
   so nth term involves ______
4. term – $2n^2$: 8, 6, 4, 2
5. Linear sequence = –2n + 10
6. So nth term is ______

Deciding if a Number is a Term

Set nth term rule equal to the number and ______. The term is in the sequence if n is an ______.

EXAMPLE

Is 37 a term in the sequence with the nth term 6n – 1?

6n – 1 = 37
6n = ...
n = 6.333...
So 37 .... in the sequence.

Other Sequences

GEOMETRIC SEQUENCE — multiply/divide previous term by ______ each time.

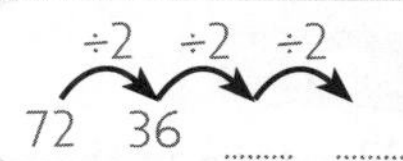

FIBONACCI-TYPE SEQUENCE — ______ previous two terms together.

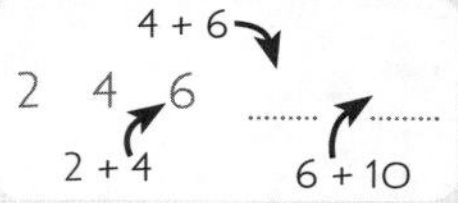

Second Go:
...../...../.....

Sequences

nth term of Linear Sequences

LINEAR SEQUENCES — increase/decrease

1. **Find**
   — this is what you
2. **Work out**
3.

EXAMPLE

Find the nth term of the sequence 7, 11, 15, 19 ...

1.
   So common difference =
2.
   so is added to each term.
3. So nth term is

nth term of Quadratic Sequences

QUADRATIC SEQUENCES — have an

1. **Find difference between**
2. **Find**
3. **Divide**
4. **Subtract n^2 term**
5. **Find**
6. **Put**

EXAMPLE

Find the nth term of the sequence 10, 14, 22, 34 ...

10 14 22 34

1.
2.
3. ,
   so nth term involves
4. term $- 2n^2$:
5. Linear sequence =
6. So nth term is

Deciding if a Number is a Term

Set nth term rule . **The term is in the sequence** .

EXAMPLE

Is 37 a term in the sequence with the nth term 6n – 1?

$6n - 1 =$ ....

$6n =$ ....

$n =$ ....

So ....

Other Sequences

GEOMETRIC SEQUENCE —

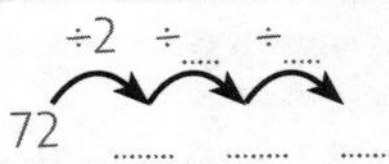

FIBONACCI-TYPE SEQUENCE —

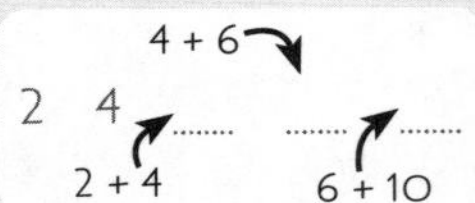

Mixed Practice Quizzes

These lovely questions will solve all of your problems. See what you remember from p.29-36, mark your answers, write down the score — you know the drill...

Quiz 1

Date: ...... / ...... / ......

1) What is the standard form for a quadratic equation?
2) What is the 12th term of the sequence with nth term $3n - 7$?
3) What are the solutions to $(3x - 2)(x + 4) = 0$?
4) What does the $\pm$ symbol mean when using the quadratic formula?
5) What is the next term in the sequence 24, 17, 10, 3, ...?
6) $x^2 + 4x - 1 = 0$ can be written $(x + 2)^2 - 5 = 0$. What are its solutions?
7) How would you find the next term in a Fibonacci-type sequence?
8) Factorise $x^2 + 6x + 5$.
9) What is the first thing you do when dividing algebraic fractions?
10) In the nth term rule for the sequence 5, 9, 17, 29, ..., what is the coefficient of n^2?

Total:

Quiz 2

Date: ...... / ...... / ......

1) When factorising a quadratic where $a = 1$, what do you do once you've found the pairs of numbers that multiply to give c?
2) What is the first step to complete the square when a isn't 1?
3) What is meant by the term 'common difference' for a linear sequence?
4) How would you find $\frac{x^2 - 4}{x^2 + 4x + 4}$ in its simplest form?
5) How do you solve a quadratic that has been factorised?
6) What type of sequence is 6, 12, 24, 48, ...?
7) What is the first step you should do to subtract algebraic fractions?
8) $x^2 - 6x + 4$ can be written $(x - 3)^2 + c$. What is the value of c?
9) What is a quadratic sequence?
10) What are the coordinates of the minimum point of $y = 2(x + \frac{5}{2})^2 - \frac{11}{2}$?

Total:

Mixed Practice Quizzes

Quiz 3

Date: ...... / ...... / ......

1) Rearrange the equation $4x^2 = 3 + 5x$ into standard quadratic form.
2) How can you use the nth term rule to work out if a number is in the sequence?
3) What is the first step you'd do to work out $\frac{1+3x}{x^2} + \frac{2x-5}{x(x+1)}$?
4) Give three reasons why you might use the quadratic formula to solve an equation.
5) What is a linear sequence?
6) What is meant by 'completing the square'?
7) How do you multiply algebraic fractions?
8) If you used the quadratic formula to solve $4x^2 = 7x - 3$, what would a, b and c be?
9) When finding the nth term of a quadratic sequence, what do you do once you've found the coefficient of n^2?
10) $x^2 + 6x - 2$ can be written as $(x + a)^2 + b$. What are the values of a and b?

Total:

Quiz 4

Date: ...... / ...... / ......

1) What is meant by 'factorising a quadratic'?
2) What is the next term in the sequence 1, 3, 4, 7, 11, ...?
3) Simplify: a) $\frac{3x^2y^3}{6x^3y}$ b) $\frac{3x+12}{(x+4)(x-1)}$
4) What is the initial bracket of the expression $x^2 - 4x + 7$ when written in completed square form?
5) State the quadratic formula.
6) What is the nth term of the sequence 5, 8, 11, 14 ...?
7) What does the adjusting number tell you about the graph of a quadratic?
8) Is 47 in the sequence with nth term rule $5n + 2$?
9) What should you use as the common denominator for $\frac{4}{x+1} - \frac{3}{x-1}$?
10) What is a geometric sequence?

Total:

Inequalities

First Go: ..... /..... /.....

Solving Inequalities

.... means **GREATER THAN** .... means **LESS THAN**

.... means **GREATER THAN OR EQUAL TO** .... means **LESS THAN OR EQUAL TO**

Solve inequalities like equations — but if you multiply/divide by a ________, flip the inequality sign.

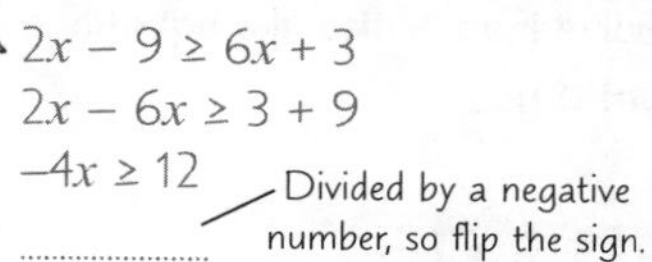

$2x - 9 \geq 6x + 3$

$2x - 6x \geq 3 + 9$

$-4x \geq 12$ — Divided by a negative number, so flip the sign.

....

EXAMPLE

Show $-2 \leq x < 5$ on the number line.

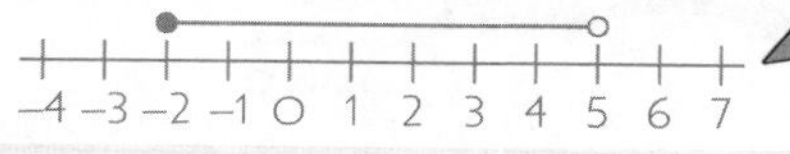

Use .... when the value is included, and .... when it's not.

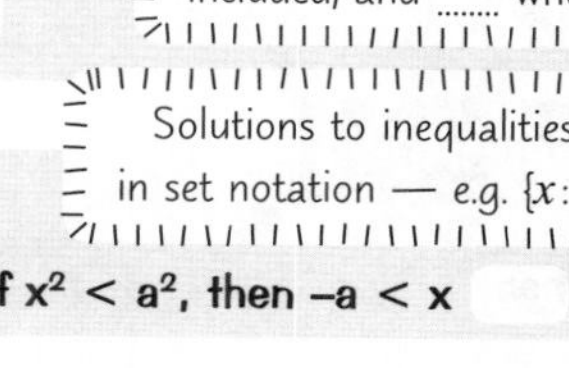

Solutions to inequalities can be given in set notation — e.g. {x: .... $\leq x \leq$ ....}.

Quadratic Inequalities

If $x^2 > a^2$, then $x > a$ or ________:

$x^2 > 36$

$x^2 = 36 \Rightarrow x = -6$ or $x = 6$

So $x <$ .... or $x >$ ....

If $x^2 < a^2$, then $-a < x$ ________:

$2x^2 \leq 98 \Rightarrow x^2 \leq 49$ — Divide both sides by 2.

$x^2 = 49 \Rightarrow x = -7$ or $x = 7$

So .... $\leq x \leq$ ....

Graphical Inequalities

1. **Convert each inequality to an ________.**
2. **Draw the ________ for each equation.**

Use a .... line if the inequality uses ≤ or ≥. Use a .... line if the inequality uses < or >.

3. **See if each inequality is ________ at a specific point, to find ________ of each line you want.**
4. **________ the region.**

EXAMPLE

Shade the region that satisfies $y < x + 4$, $y \leq 1 - 2x$ and $y > -1$.

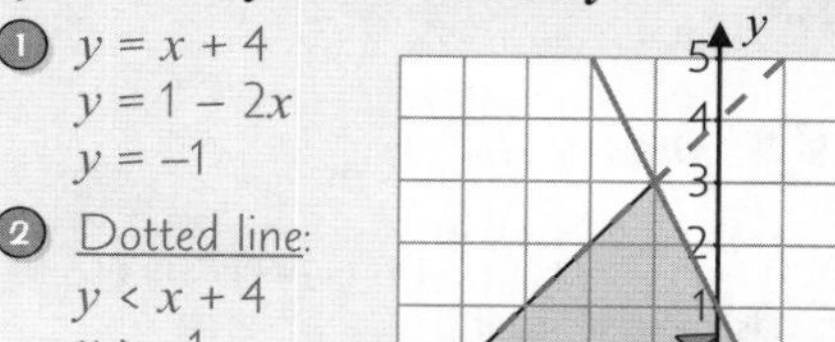

1. $y = x + 4$
   $y = 1 - 2x$
   $y = -1$
2. Dotted line:
   $y < x + 4$
   $y > -1$
   Solid line:
   $y \leq 1 - 2x$

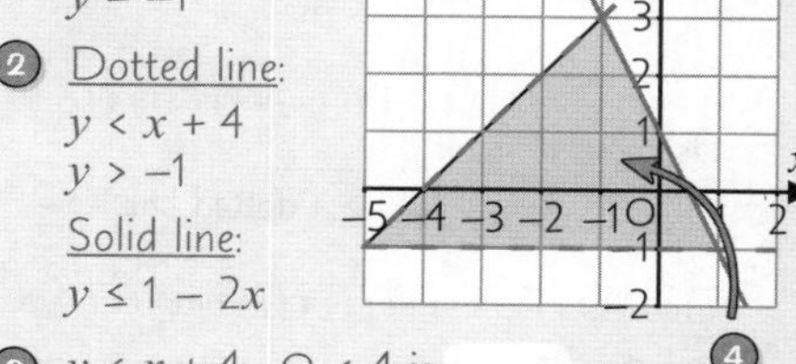

3. $y < x + 4$: $0 < 4$ is ________, so (0, 0) is on ________ of line.
   $y \leq 1 - 2x$: $0 \leq 1$ is ________, so (0, 0) is on ________ of line.
   $y > -1$: $0 > -1$ is ________, so (0, 0) is on ________ of line.

Second Go: ...../...../.....

Inequalities

Solving Inequalities

$>$ means .... $<$ means ....

...... means **GREATER THAN OR EQUAL TO** ...... means **LESS THAN OR EQUAL TO**

Solve inequalities like equations — but if you ....

$2x - 9 \geq 6x + 3$

.... $\geq$ ....

.... $\geq$ ....

.... $\leq$ .... — Divided by a negative number, so flip the sign.

EXAMPLE

Show $-2 \leq x < 5$ on the number line.

–4 –3 –2 –1 0 1 2 3 4 5 6 7

Use ● ...., and ○ ....

Solutions to inequalities can be given in .... — e.g. {........................}.

Quadratic Inequalities

If $x^2 > a^2$, then .... **:**

$x^2 > 36$

.... $\Rightarrow$ .... or ....

So ....

If $x^2 < a^2$, then .... **:**

$2x^2 \leq 98 \Rightarrow$ .... — Divide both sides by 2.

.... $\Rightarrow$ .... or ....

So ....

Graphical Inequalities

1. **Convert**
2. **Draw**

Use a solid line if .... Use a dotted line if ....

3. **See if each inequality is true**
4. ....

EXAMPLE

Shade the region that satisfies $y < x + 4$, $y \leq 1 - 2x$ and $y > -1$.

1. ....
2. Dotted line: ....

Solid line: ....

3. $y < x + 4$: ...., so (0, 0) ....

$y \leq 1 - 2x$: ...., so (0, 0) ....

$y > -1$: ...., so (0, 0) ....

4

Iterative Methods

First Go: ..... /..... /.....

Using Iterative Methods

ITERATIVE METHODS — ____ ____ to get closer to the actual solution.

They're used when equations are ____ to solve.

You usually keep putting the .... you've just found .... into the calculation.

For an equation that equals 0:

Substitute ____ into the equation.

If the sign ____, there's a ____ between the two numbers.

Decimal Search Method

1. Substitute 1 d.p. values of x within the ____ until the ____.
2. Substitute values of x with 2 d.p. until the ____ again.
3. ____ until values either side of the sign change are the ____ when rounded to required degree of accuracy.

EXAMPLE

The equation $x^3 - 5x - 1 = 0$ has a solution between $x = 0$ and $x = -1$. Find this solution to 1 d.p.

	x	$x^3 - 5x - 1$	Sign
	0	−1	
	−0.1	−0.501	
1	−0.2	−0.008	
	−0.3	0.473	
2	−0.20	−0.008	
	−0.21	0.040739	

3. Both −0.20 and −0.21 round to ____ to 1 d.p., so the solution is ____.

Iteration Machines

EXAMPLE

Use the iteration machine below to find a solution to $x^3 - 7x - 2 = 0$ to 1 d.p. Use the starting value $x_0 = 2$.

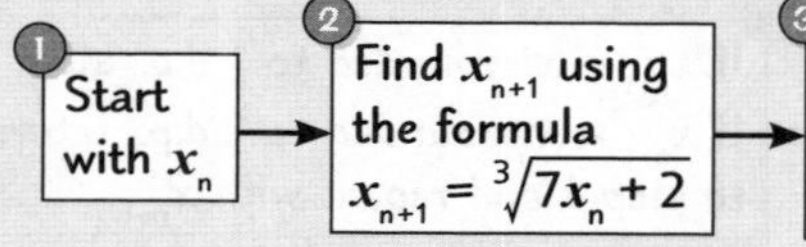

Follow the instructions in the iteration machine:

$x_0 = 2$

$x_1 = 2.519...$ ____ x_0 to 1 d.p.

$x_2 = 2.697...$ ____ x_1 to 1 d.p.

$x_3 = 2.753...$ ____ x_2 to 1 d.p.

$x_4 = 2.771...$ ____ x_3 to 1 d.p.

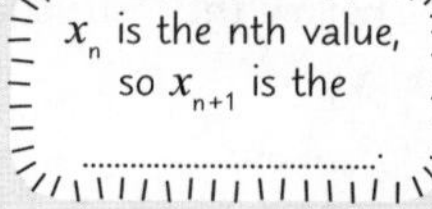

x_3 and x_4 both round to ____, so the solution is ____ (1 d.p.).

Second Go: ...../...../.....

Iterative Methods

Using Iterative Methods

ITERATIVE METHODS —

They're used when equations are

For an equation that :

Substitute → **If the sign changes,**

Decimal Search Method

1. **Substitute 1 d.p. values of x**
2. **Substitute values of x with 2 d.p.**
3. **Repeat until values either side of the sign change**

EXAMPLE

The equation $x^3 - 5x - 1 = 0$ has a solution between $x = 0$ and $x = -1$. Find this solution to 1 d.p.

	x	$x^3 - 5x - 1$	Sign
	0		
	−0.1		
1	−0.2		
	−0.3		
2	−0.20		
	−0.21		

3 Both

to 1 d.p., so the solution is .

Iteration Machines

EXAMPLE

Use the iteration machine below to find a solution to $x^3 - 7x - 2 = 0$ to 1 d.p. Use the starting value $x_0 = 2$.

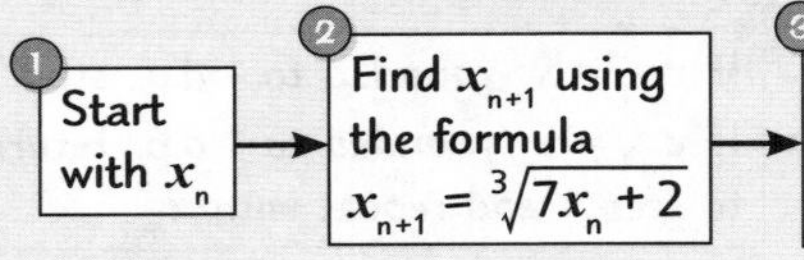

Follow the instructions in the iteration machine:

$x_0 =$

$x_1 =$ to 1 d.p.

$x_2 =$ to 1 d.p.

$x_3 =$ to 1 d.p.

$x_4 =$ to 1 d.p.

both round to , so the solution is (1 d.p.).

x_n is the ...., so x_{n+1} is the ....

Simultaneous Equations

First Go: ..... /..... /.....

Six Steps for Easy Ones

When both equations are :

1. Rearrange into the form .
2. Match up the for one of the variables.
3. to get rid of a variable.
4. Solve the equation.
5. Substitute the value back into the equation.
6. your answer works.

EXAMPLE

Solve the simultaneous equations
$5 - 2x = 3y$ and $5x + 4 = -2y$

1. (1) (2) ← Label your equations.
2. $(1) \times 5$: $10x + 15y = 25$ (3)
   $(2) \times 2$: $10x + 4y = -8$ (4)
3. $(3) - (4)$:
4. $11y = 33 \Rightarrow y =$
5. Sub $y = 3$ into (1): $2x + (3 \times 3) = 5$
   $\Rightarrow 2x = 5 - 9 \Rightarrow 2x = -4 \Rightarrow x =$
6. Sub x and y into (2):
   $(5 \times -2) + (2 \times 3) = -4$
   So the solution is , .

Seven Steps for Tricky Ones

When one equation is :

1. Rearrange one equation so a is by itself.
2. Substitute the rearranged equation into the .
3. Rearrange and solve.
4. Substitute into one of the equations.
5. Substitute into the same equation.
6. both pairs of solutions work.
7. Write out both pairs of solutions clearly.

EXAMPLE

Solve the simultaneous equations
$3x - y = 5$ and $3x^2 - y = 11$

1. (1) (2)
2. $3x - (3x^2 - 11) = 5$ (3)
3. $3x^2 - 3x - 6 = 0$
   $3(x - 2)(x + 1) = 0$
   So $x - 2 = 0 \Rightarrow$
   or $x + 1 = 0 \Rightarrow$

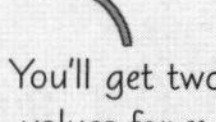

4. Sub into (1):
   $6 - y = 5$, so
5. Sub into (1):
   $-3 - y = 5$, so
6. Sub both x-values into (2):
   $x = 2$: $y = (3 \times 2^2) - 11 =$
   $x = -1$: $y = (3 \times (-1)^2) - 11 =$
7. and

Second Go:
...../...../.....

Simultaneous Equations

Six Steps for Easy Ones

When :

1. Rearrange
2. Match up
3. Add or subtract
4.
5. Substitute
6. Check

EXAMPLE

Solve the simultaneous equations
$5 - 2x = 3y$ and $5x + 4 = -2y$

1. (1) ← Label your equations.
   (2)
2. (1) × 5: (3)
   (2) × 2: (4)
3. (3) − (4):
4. ⇒
5. Sub into (1):
   ⇒ ⇒ ⇒
6. Sub x and y into (2):

Seven Steps for Tricky Ones

When :

1. Rearrange
2. Substitute
3.
4. Substitute
5. Substitute
6. Check
7. Write out

EXAMPLE

Solve the simultaneous equations
$3x - y = 5$ and $3x^2 - y = 11$

1. (1)
   (2)
2. (3)
3. = 0
   = 0
   So ⇒
   or ⇒ ← You'll get two values for x.
4. Sub into (1):
5. Sub into (1):
6. Sub into (2):
7.

Proof

First Go: ..... /..... /.....

Five Facts for Algebraic Proof

'n' stands for any ....

1. Even Numbers — can be written as ____.
2. Odd Numbers — can be written as ____.
3. Multiples — can be written as ____ (e.g. write multiples of 3 as 3n).
4. Consecutive Numbers — can be written as ____, ____, ____, etc.
5. The sum, difference or product of integers is always an ____.

Proof Examples

EXAMPLE

Show that the product of two odd numbers is always odd.

Odd numbers: $2a + 1$ and $2b + 1$.

$(2a + 1)(2b + 1) = 4ab + 2a + 2b + 1$
$= 2(2ab + a + b) + 1$

This can be written as ____, where ____, so it must be odd.

EXAMPLE

Prove $(n - 4)^2 - (n + 1)^2 \equiv -5(2n - 3)$.

$(n - 4)^2 - (n + 1)^2$
$\equiv (n^2 - 8n + 16) - (n^2 + 2n + 1)$
$\equiv n^2 - 8n + 16 - n^2 - 2n - 1$
$\equiv$ ____
$\equiv$ ____

The identity symbol '≡' means this is true for ....

EXAMPLE

Given $\angle CED = x$, show that $\angle CAB = \frac{1}{2}(90° + x)$.

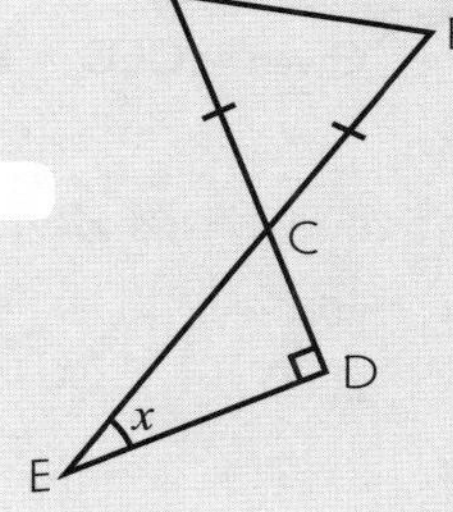

$x + 90° + \angle ECD = 180°$, so $\angle ECD =$ ____

$\angle ECD$ and $\angle ACB$ are vertically opposite, so $\angle ACB =$ ____

Triangle ABC is ____, so $\angle CAB = \angle ABC$

$2\angle CAB + (90° - x) = 180°$

$2\angle CAB = 90° + x$

$\angle CAB = \frac{1}{2}(90° + x)$

This is a .... proof.

Disproof by Counter-example

Prove that a statement is false by finding a ____.
Keep trying numbers until you find one that ____.

EXAMPLE

Disprove the statement: "The sum of two square numbers is always odd."

$1 + 4 = 5$ (____) $4 + 9 = 13$ (____)

$1 + 9 = 10$ (____) so the statement is ____.

Second Go:
...../...../.....

Proof

Five Facts for Algebraic Proof

1. **Even Numbers —** .
2. **Odd Numbers —** .
3. **Multiples —**
   (e.g. write multiples of 3 as).
4. **Consecutive Numbers —** .
5. **The , or of integers is** .

'n' stands for
....

Proof Examples

EXAMPLE

Show that the product of two odd numbers is always odd.

Odd numbers: and

( )( )

=

=

This can be written as

, so it must be .

EXAMPLE

Prove $(n - 4)^2 - (n + 1)^2 \equiv -5(2n - 3)$.

$(n - 4)^2 - (n + 1)^2$

$\equiv$ ( ) − ( )

$\equiv$

$\equiv$

$\equiv$

The identity symbol '.........' means this is
....
....

EXAMPLE

Given ∠CED = x, show that ∠CAB = $\frac{1}{2}$(90° + x).

= 180°, so ∠ECD =

∠ECD and ∠ACB are , so ∠ACB =

Triangle ABC is , so

= 180°

This is ....

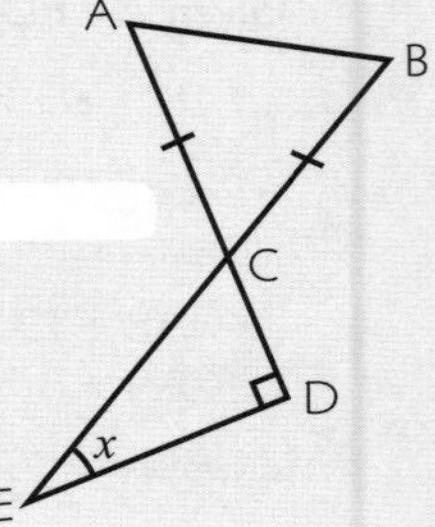

Disproof by Counter-example

Prove that a statement is

Keep trying numbers

EXAMPLE

Disprove the statement: "The sum of two square numbers is always odd."

(odd) (odd)

(even) so the statement is false.

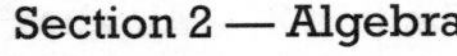

Functions

First Go: ..... /..... /.....

Evaluating Functions

FUNCTION — takes an ______ , processes it, ______ a value.

They're usually written like:

$f(x) = (x + 2)^2 - 5$

This means "take a value of x, ______ , square it, then ______ ".

______ functions by just substituting in the value of x.

$f(-4) = (-4 + 2)^2 - 5$
$= \dots\dots\dots = \dots\dots$

Functions can also be written like $f : x \rightarrow$ ....

Composite Functions

COMPOSITE FUNCTION — two functions combined into a ______ .

fg(x) → put ______ into ______

gf(x) → put ______ into ______

Three steps for composite functions:

1. **Write composite function with ______ .**
2. **Replace first function with ______ .**
3. **Substitute it into ______ .**

EXAMPLE

$f(x) = 4x - 1$ and $g(x) = \frac{3x}{2}$.

a) Find $fg(x)$.

$f(g(x))$ ① $= f($ ______ $)$ ② $= 4 \times \frac{3x}{2} - 1$ ③
$=$ ______

b) Find $gf(x)$.

$g(f(x))$ ① $= g($ ______ $)$ ② $= \frac{3(4x-1)}{2}$ ③
$= \frac{12x-3}{2}$
$=$ ______

In general, $fg(x)$ .... $gf(x)$.

Inverse Functions

INVERSE FUNCTION, $f^{-1}(x)$ — a function that ______ f(x).

Three steps for inverse functions:

1. **Write the equation ______ .**
2. **Make ______ the subject.**
3. **Replace y with ______ .**

EXAMPLE

Given $f(x) = 7x - 11$, find $f^{-1}(x)$.

① ______ ← Replace $f(x)$ with x and x with y.

② $7y = x + 11$
$y = \frac{x + 11}{7}$

③ ______

Check it .... the function: $f(2) = 3$, and $f^{-1}(3) = 2$ ✓

Second Go:
...../...../.....

Functions

Evaluating Functions

FUNCTION — .

They're usually written like:

$f(x) = (x + 2)^2 - 5$

This means " ".

Functions can also be written like .

Evaluate functions by

$f(-4) =$ ....

$=$ .... $=$ .....

Composite Functions

COMPOSITE FUNCTION —

fg(x) →

gf(x) →

Three steps for composite functions:

1. **Write**
2. **Replace**
3. **Substitute**

EXAMPLE

$f(x) = 4x - 1$ and $g(x) = \frac{3x}{2}$.

a) Find fg(x).

$f(g(x)) =$ ① $=$ ② $=$ ③

b) Find gf(x).

$g(f(x)) =$ ① $=$ ② $=$ ③ $=$

In general, .

Inverse Functions

INVERSE FUNCTION, $f^{-1}(x)$ —

Three steps for inverse functions:

1. **Write** .
2. **Make** .
3. **Replace** .

EXAMPLE

Given $f(x) = 7x - 11$, find $f^{-1}(x)$.

①

②

Replace $f(x)$ with x and x with y.

$y =$

③

Check it .... the function:

$f(2) =$ ....., and $f^{-1}(3) =$ ..... ✓

Mixed Practice Quizzes

Prove your knowledge has no equal by having a go at these questions, covering everything on p.39-48. Mark your answers, and write your score in the box.

Quiz 1

Date: ...... / ...... / ......

1) How many pairs of solutions will you get when solving a linear and a quadratic equation simultaneously?
2) What is a function?
3) List the integers that satisfy the inequality $-3 < x \leq 1$.
4) How would you show that $6(2n + 1) - 2(4n - 1)$ is always a multiple of 4?
5) If $x^2 \geq a^2$, what are the possible values of x?
6) What is the first step you would do to use the decimal search method?
7) If $f(x) = x^2$ and $g(x) = x + 1$, what is: a) $fg(x)$? b) $gf(x)$?
8) How could you write an odd number in terms of an integer n?
9) Would the inequality $4x + 3y > 2$ have a solid or dotted line on a graph?
10) For a function $g(x)$, what is $g^{-1}(x)$?

Total:

Quiz 2

Date: ...... / ...... / ......

1) Solve $3x - 7 \geq 2$.
2) How could you check your solution to a pair of simultaneous equations?
3) What is disproof by counter-example?
4) How do you evaluate a function $f(x)$?
5) What do you need to do to the inequality sign when you multiply or divide by a negative number?
6) When would you use an iterative method to solve an equation?
7) How would you prove that the sum of two odd numbers is always even?
8) What lines would you draw to show the inequalities $y > 3x - 2$ and $y \leq 6 - 2x$ on a graph?
9) Given $f(x) = 5x + 2$, what is $f^{-1}(x)$?
10) What is meant by x_n in an iterative formula?

Total:

Mixed Practice Quizzes

Quiz 3

Date: ...... / ...... / ......

1) Write "take a value of x, multiply it by 3, then add 7" as a function.
2) Would $x > 6$ have a solid or hollow circle when shown on a number line?
3) How could you write an even number in terms of an integer n?
4) What is an iterative method?
5) For functions f(x) and g(x), how would you find fg(x)?
6) What is the first step you would take to solve simultaneous equations when one of them is quadratic?
7) What is an inverse function?
8) Solve the inequality $x^2 - 8 < 28$.
9) For an equation $f(x) = 0$: $f(1.1) = 2$, $f(1.15) = 0.75$ and $f(1.2) = -2$. What is the root of the equation to the nearest 1 decimal place?
10) Disprove the statement: "The sum of two primes is never prime."

Total:

Quiz 4

Date: ...... / ...... / ......

1) How could you write two consecutive numbers in terms of an integer n?
2) What is the range of solutions for $3x - 8 < 5x + 6$?
3) $f(x) = x^3 - 7x + 3$. a) What is f(2)? b) What is f(–1)?
4) What is an iteration machine?
5) What is the first thing you would do to solve the equations $2x - 3y = 8$ and $5x + 4y = -3$ simultaneously?
6) What does the symbol $\equiv$ mean?
7) What is a composite function?
8) How do you find out which side of each line you want when plotting graphical inequalities?
9) Show that $x^4 + 9x - 7 = 0$ has a solution between $x = 0$ and $x = 1$.
10) How can you check that an inverse function is correct?

Total:

Straight-Line Graphs

First Go: ..... /..... /.....

Straight-Line Equations

'x = a' is a line through 'a' on the x-axis (e.g. x = –3)

'y = a' is a line through 'a' on the y-axis (e.g. y = –1)

The x-axis is $y =$ ..... and the y-axis is $x =$ .....

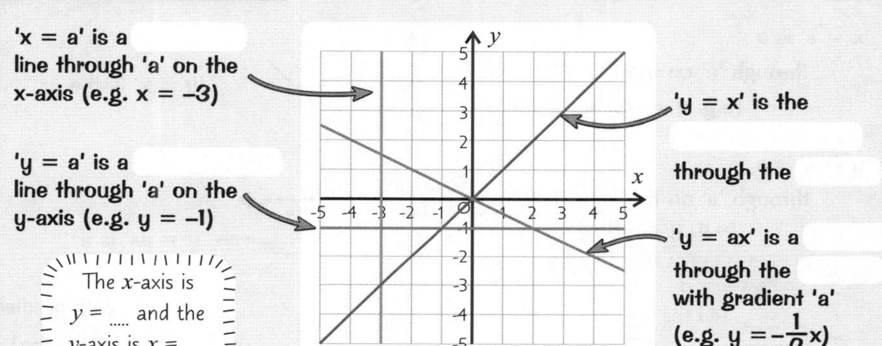

'y = x' is the through the

'y = ax' is a through the with gradient 'a' (e.g. $y = -\frac{1}{2}x$)

Equations of Straight-Line Graphs

GRADIENT — of a line.

$$\text{Gradient} = \frac{\text{change in ......}}{\text{change in ......}}$$

1. Use any two points on the line to find the , 'm'.
2. Find the , 'c'.
3. Write equation as .

EXAMPLE

1. $m = \frac{4}{6} = \frac{......}{......}$
2. $c =$ ......

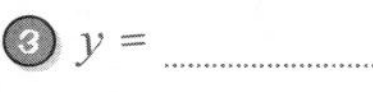

3. $y =$ ....

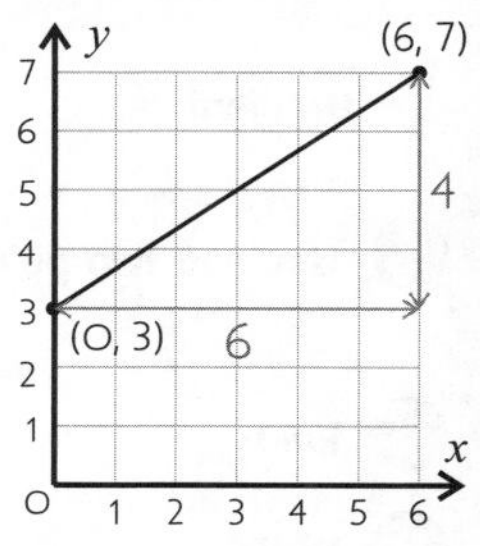

Equation of a Line Through Two Points

1. Use both points to find .
2. Substitute one point into .
3. Rearrange to find ' '.
4. Write equation as .

EXAMPLE

Find the equation of the straight line that passes through (–2, 12) and (4, –6).

1. $m = \frac{-6 - 12}{4 - (-2)} = \frac{\quad}{\quad} = \quad$
2. Sub in (4, –6):
   $\quad = -3(\quad) + c \Rightarrow \quad = \quad + c$
3. $c = \quad = \quad$
4. $y = \quad$

Second Go:
...../...../.....

Straight-Line Graphs

Straight-Line Equations

'x = a' is a
through 'a' on the
(e.g. x = –3)

'y = a' is a
through 'a' on the
(e.g. y = –1)

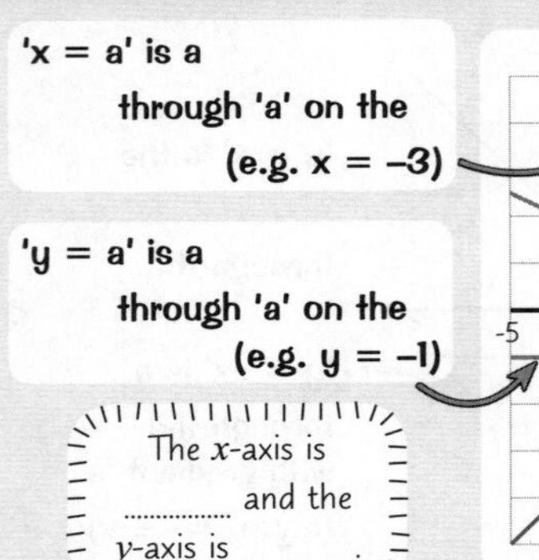

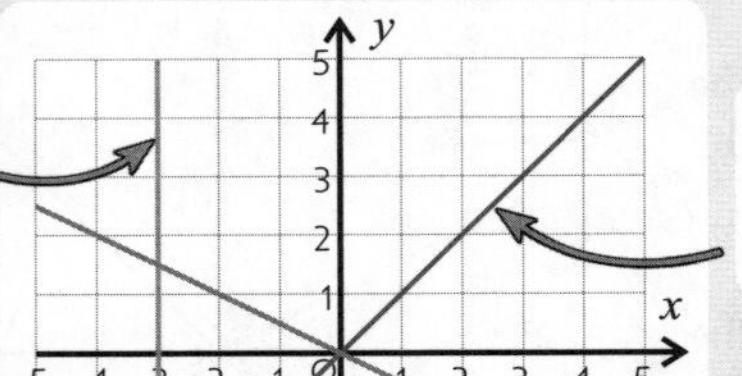

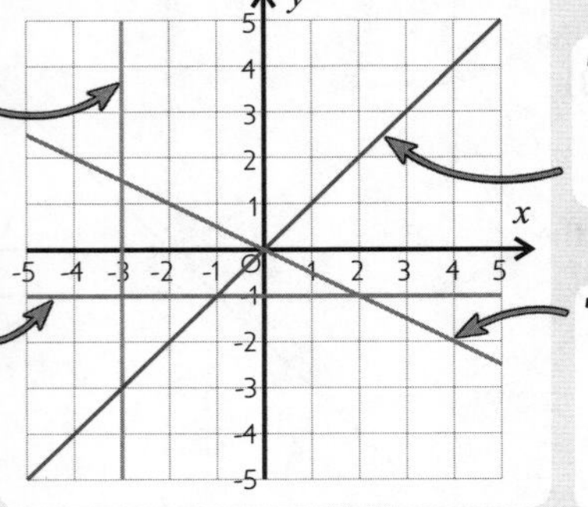

'y = x' is the

'y = ax' is a

with gradient
(e.g. $y = -\frac{1}{2}x$)

Equations of Straight-Line Graphs

GRADIENT —

Gradient =

1. Use any two points
2. Find
3. Write

EXAMPLE

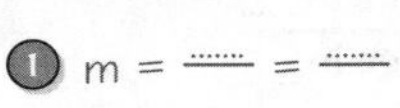

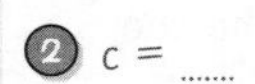

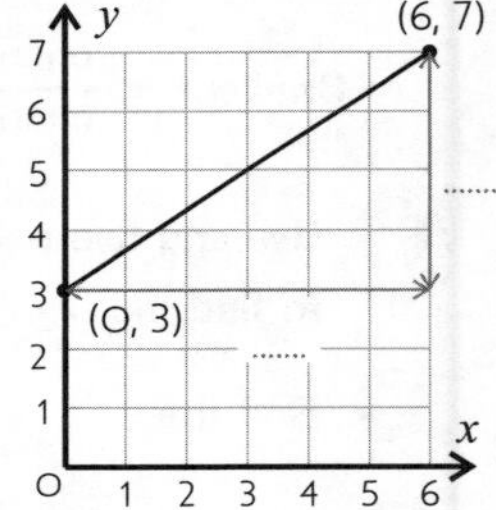

Equation of a Line Through Two Points

1. Use
2. Substitute
3. Rearrange
4. Write

EXAMPLE

Find the equation of the straight line that passes through (–2, 12) and (4, –6).

1. m = = =
2. Sub in (4, –6):
   ⇒
3. c = =
4.

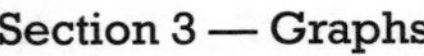

Drawing Straight-Line Graphs

First Go: ...../...../.....

'Table of 3 Values' Method

1. Draw a table with three ____.
2. Put the ____ into the equation and work out the ____.
3. ____ the points and ____ a line through them.

EXAMPLE

Draw the graph $y = -2x + 3$ for values of x from 0 to 4.

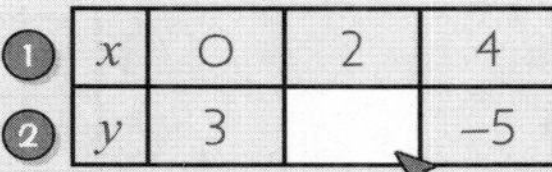

① x	0	2	4
② y	3		–5

E.g. when $x = 2$,
$y = -2(\ \ ) + 3$
$= \ \ + \ \ = \ \ $

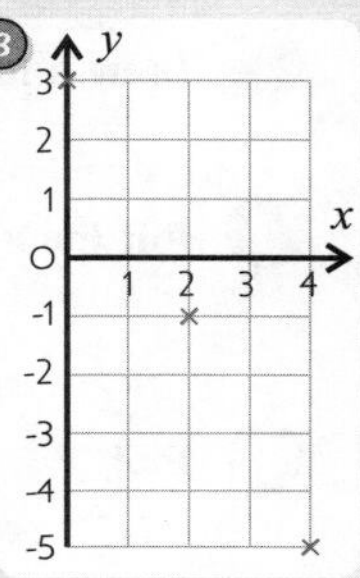

Using y = mx + c

1. Rearrange into the form ____.
2. Put a ____ on the y-axis at the value of ____.
3. Use ____ to go ____ and ____ an appropriate number of units. Make a ____ and ____.
4. Draw a ____ through the dots.
5. Check ____ looks correct.

EXAMPLE

Draw the graph of $3y = x + 6$.

① $3y = x + 6 \Rightarrow y =$ ____

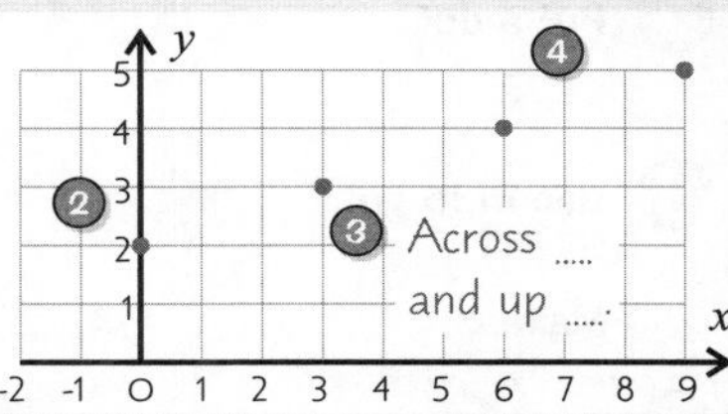

⑤ A gradient of ____ is gentle and increases from left to right. ✓

'x = 0, y = 0' Method

1. Set x = 0 and ____.
2. Set y = 0 and ____.
3. Mark and label ____ ____. Draw a line through them.

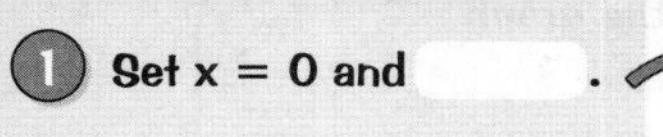

EXAMPLE

Sketch the graph of $y = 2x - 3$.

① When $x = 0$,
$y = 2(\ \ ) - 3$
$= \ \ $

② When $y = 0$,
$\ \ = 2x - 3$
$\Rightarrow x = \ \ $

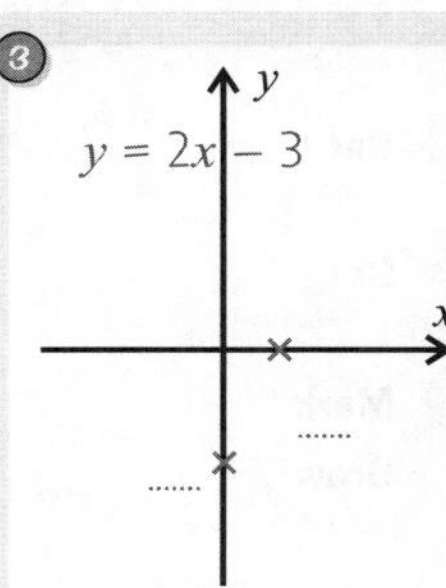

Second Go: ...../...../.....

Drawing Straight-Line Graphs

'Table of 3 Values' Method

1. Draw a table
2. Put the x-values
3. Plot

EXAMPLE

Draw the graph $y = -2x + 3$ for values of x from 0 to 4.

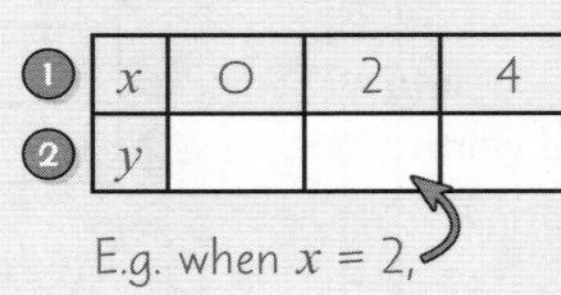

① x	0	2	4
② y			

E.g. when $x = 2$,

$y =$

$=$ $=$

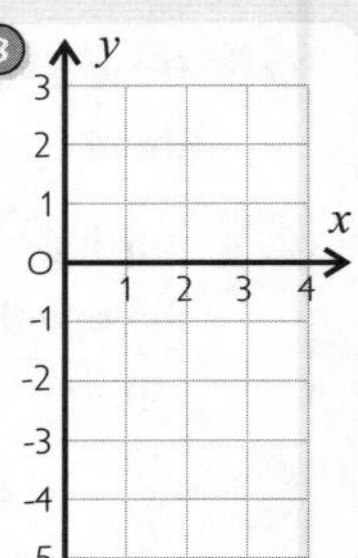

Using y = mx + c

1. Rearrange
2. Put a dot
3. Use m to go
   Make
4. Draw
5. Check

EXAMPLE

Draw the graph of $3y = x + 6$.

① $3y = x + 6 \Rightarrow$

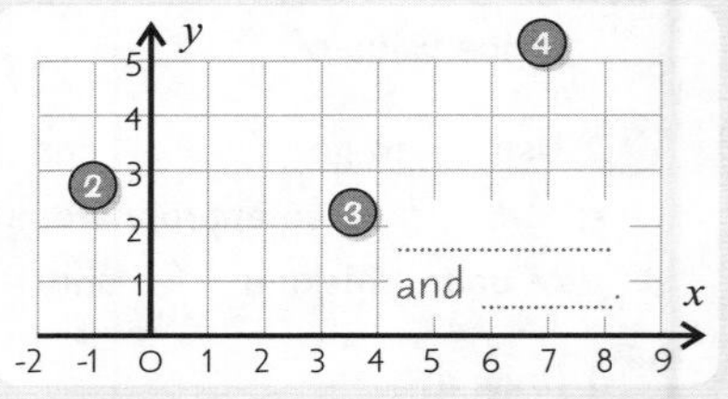

⑤ A gradient of is and from left to right. ✓

'x = 0, y = 0' Method

1. Set
2. Set
3. Mark
   Draw

EXAMPLE

Sketch the graph of $y = 2x - 3$.

① When $x = 0$,

$y =$

$=$

② When $y = 0$,

$=$

$\Rightarrow x =$

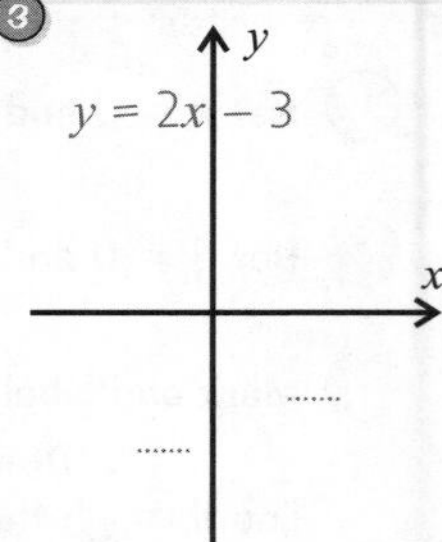

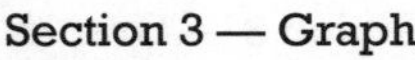

Working with Straight-Line Graphs

First Go: ..... /..... /.....

Parallel Line Gradients

Parallel lines have the gradient.
— i.e. they have the m value.

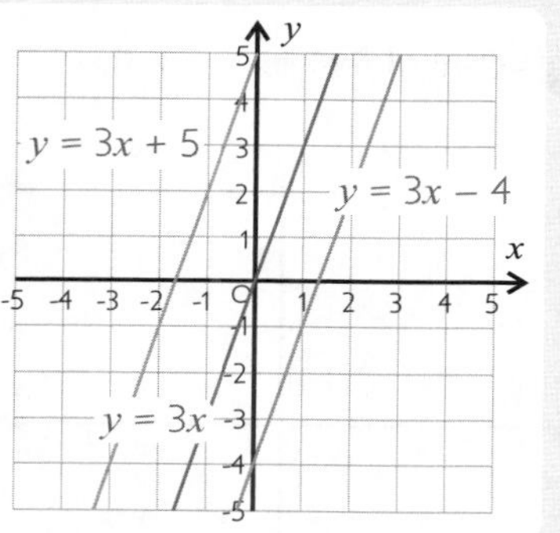

Perpendicular Line Gradients

Perpendicular lines cross at .
Their gradients multiply together to give .
If gradient of first line = m,
then gradient of second line =

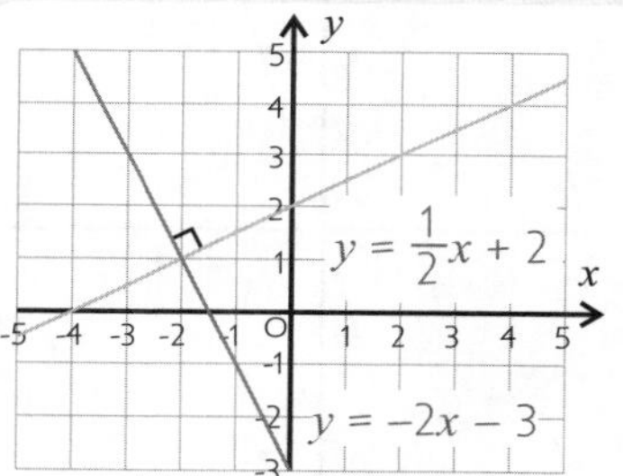

Two Steps to Find the Midpoint of a Line Segment

1. Add the
of the
and divide by .
2. Add the
of the
and divide by .

EXAMPLE

Point A has coordinates (–8, 2)
and Point B has coordinates (6, 10).
Find the coordinates of the midpoint of AB.

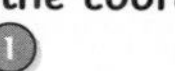

$\left(\frac{\dots + \dots}{2}, \frac{\dots + \dots}{2}\right) = \left(\frac{\quad}{\quad}, \frac{\quad}{\quad}\right) = \dots$

Using Ratios to Find Coordinates

1. Find the
between the x-coordinates
and the y-coordinates.
2. Use the to find the difference
between a and
the point you want to find.
3. the differences
to the .

EXAMPLE

R = (–3, –7) and S = (2, 3).
T lies on RS, so that RT:TS = 2:3.
Find the coordinates of T.

1. Difference between x-coordinates =
Difference between y-coordinates =
2. T is $\frac{2}{2+3}$ = along RS from R, so
x: $\frac{2}{5} \times$ = y: $\frac{2}{5} \times$ =
3. T = (–3 + , –7 +) =

Second Go: ...../...../.....

Working with Straight-Line Graphs

Parallel Line Gradients

Parallel lines

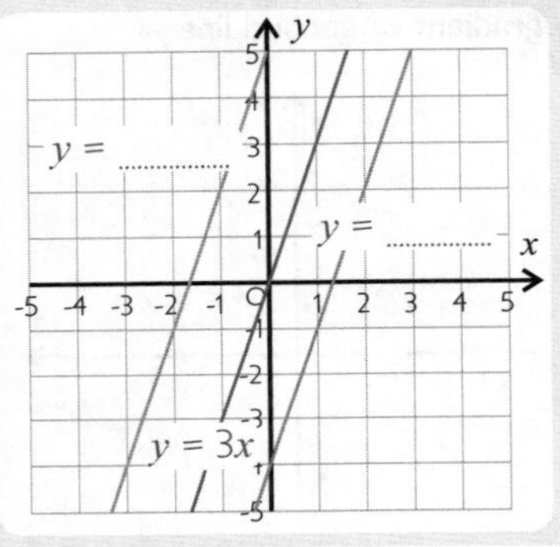

Perpendicular Line Gradients

Perpendicular lines

If gradient of first line = m, then

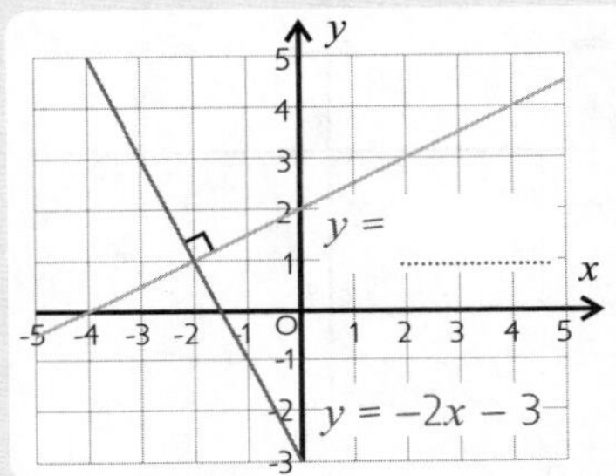

Two Steps to Find the Midpoint of a Line Segment

1. **Add**

2. **Add**

EXAMPLE

Point A has coordinates (–8, 2) and Point B has coordinates (6, 10). Find the coordinates of the midpoint of AB.

① ②

$\left(\frac{\quad}{\quad}, \frac{\quad}{\quad}\right) = \left(\frac{\quad}{\quad}, \frac{\quad}{\quad}\right) =$ ....

Using Ratios to Find Coordinates

1. **Find the difference**

2. **Use the ratio**

3. **Add**

EXAMPLE

R = (–3, –7) and S = (2, 3). T lies on RS, so that RT:TS = 2:3. Find the coordinates of T.

① Difference between x-coordinates =
Difference between y-coordinates =

② T is $\frac{\quad}{\quad} = \frac{\quad}{\quad}$ along RS from R, so

x: y:

③ T = =

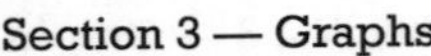

Quadratic and Cubic Graphs

First Go: ...../...../.....

Quadratic Graphs

A quadratic graph (y = + +) has a symmetrical bucket shape.

Three steps to plot a quadratic graph:

1. Substitute the into the equation to find .
2. the points.
3. Join the points with a .

The coefficient of x^2 is positive, so the curve is ....

EXAMPLE

Plot the graph of $y = x^2 + 2x - 1$.

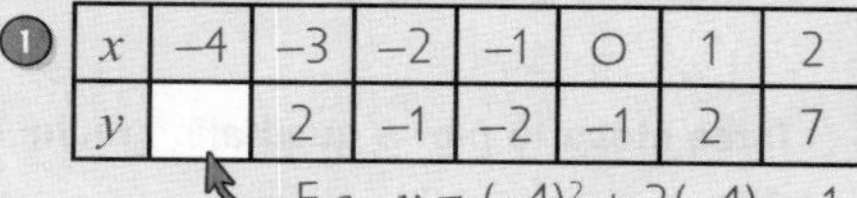

x	–4	–3	–2	–1	0	1	2
y		2	–1	–2	–1	2	7

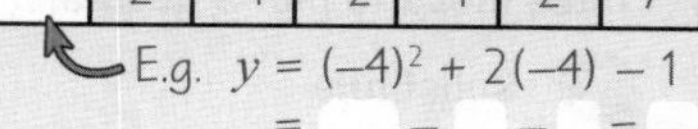

E.g. $y = (-4)^2 + 2(-4) - 1$
= – – =

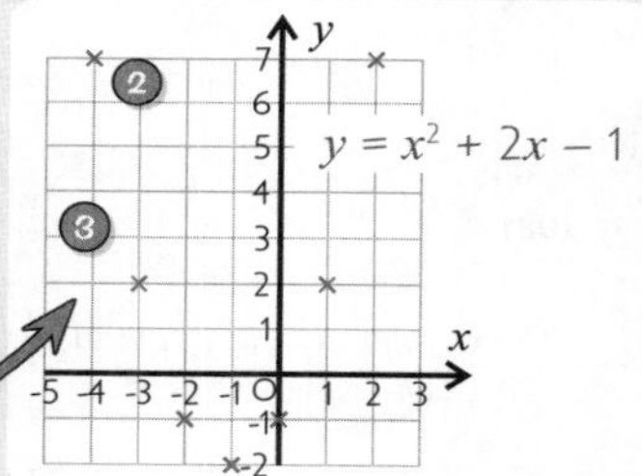

Sketching Quadratics

1. Find the .
2. Use to find the x-coordinate of the .
3. Substitute into the equation to find .
4. Sketch and graph.

EXAMPLE

Sketch the graph of $y = -x^2 - 2x + 3$.

1. $-x^2 - 2x + 3 = -(x + 3)(x - 1)$
So $x =$ and $x =$
2. $x = \frac{ + }{2} =$
3. $y = -()^2 - 2() + 3$
=

The coefficient of x^2 is negative, so the curve is ....

4.

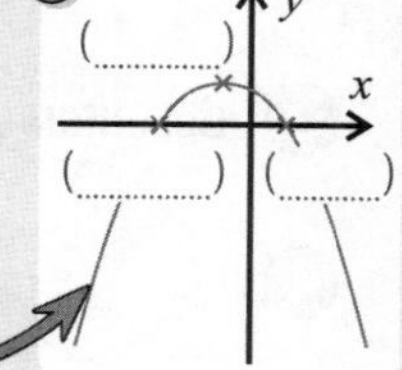

Cubic Graphs

A cubic graph (y = + + +) has a wiggle in the middle.

$+x^3$ graphs go up from : $-x^3$ graphs go down from :

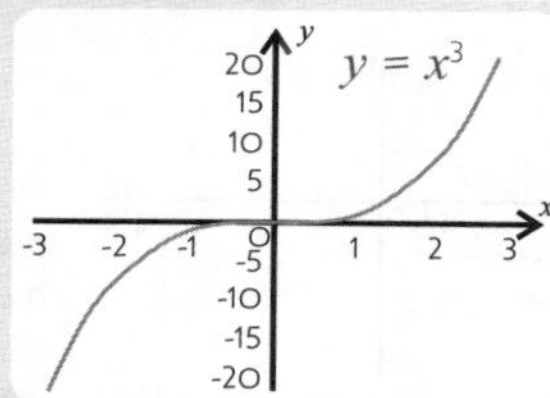

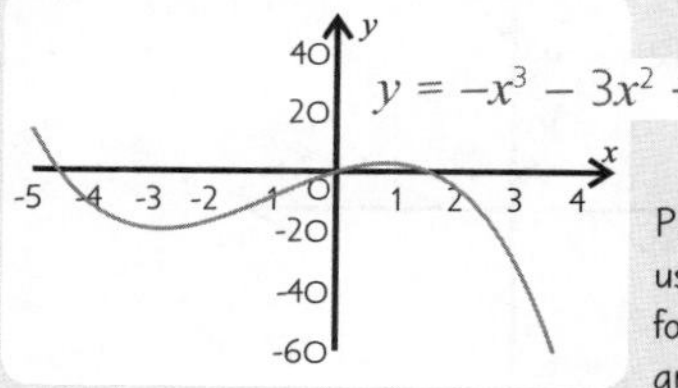

Plot cubic graphs using the steps for quadratic graphs above.

Second Go:
...../...../.....

Quadratic and Cubic Graphs

Quadratic Graphs

A quadratic graph (y =)

Three steps to plot a quadratic graph:

1. Substitute
2.
3. Join

The coefficient of x^2 is ....,
....

EXAMPLE

Plot the graph of $y = x^2 + 2x - 1$.

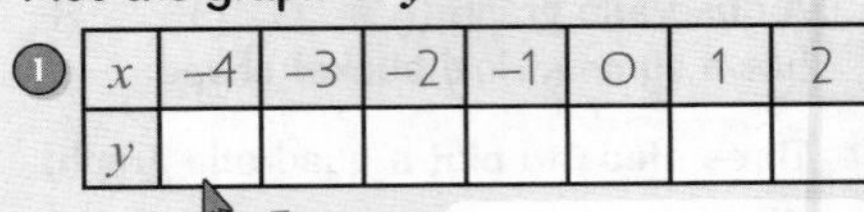

x	−4	−3	−2	−1	0	1	2
y							

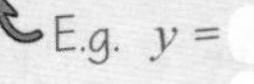

E.g. $y =$
$=$ $=$

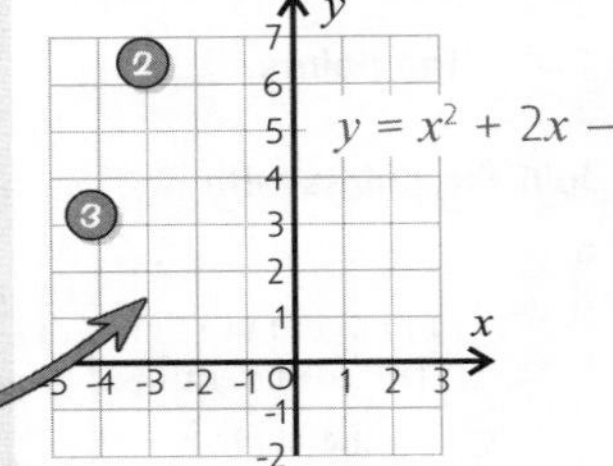

Sketching Quadratics

1. Find
2. Use symmetry to
3. Substitute
4.

EXAMPLE

Sketch the graph of $y = -x^2 - 2x + 3$.

1. $-x^2 - 2x + 3 = -($ $)($ $)$
   So $x =$ and $x =$
2. $x = \frac{\quad}{\quad} =$
3. $y =$
   $=$

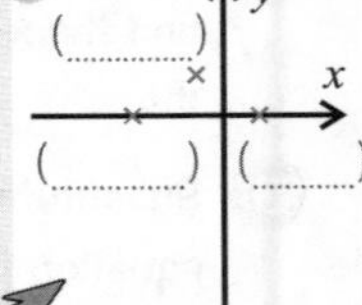

The coefficient of x^2 is ....,
....

Cubic Graphs

A cubic graph (y =)

$+x^3$ graphs go

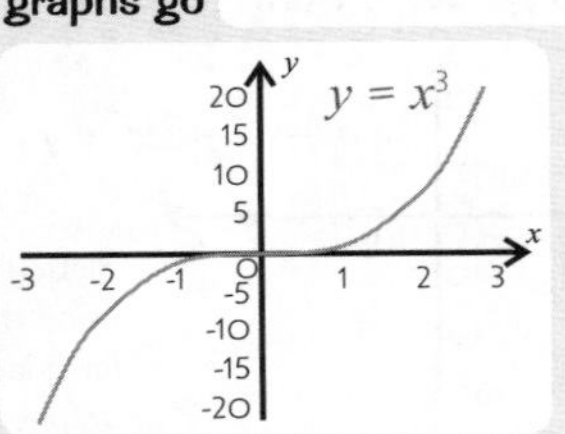

$-x^3$ graphs go

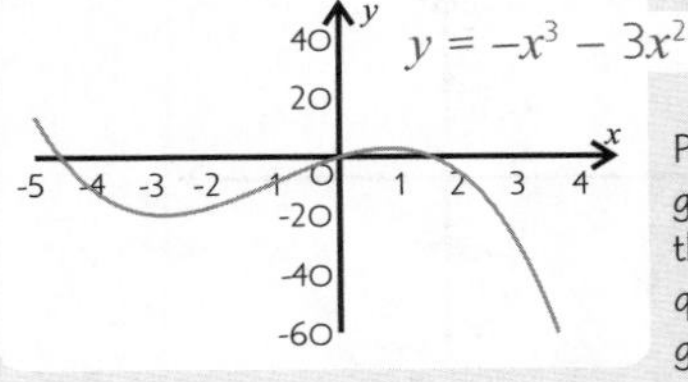

Plot cubic graphs using the steps for quadratic graphs above.

Mixed Practice Quizzes

You've plotted a straight-line course to these quizzes on p.51-58.
When you've given them a go, mark each test and add up your score.

Quiz 1

Date: ...... / ...... / ......

1) Give the standard format of the equation of a cubic graph.
2) True or false? Perpendicular lines have the same gradient.
3) Describe the graph of $x = 2$.
4) If a quadratic graph has a negative x^2 coefficient, will the graph be n-shaped or u-shaped?
5) How do you find the midpoint of a line segment?
6) What is the gradient of the line passing through (1, 0) and (3, 4)?
7) Rearrange $2x + y = 5$ into the form $y = mx + c$.
8) Describe how to plot a quadratic graph from its equation.
9) The graph of $y = 2x^2 - 4x$ passes through the points (–1, 6) and (3, 6). What is the turning point of $y = 2x^2 - 4x$?
10) What is the equation of the y-axis?

Total:

Quiz 2

Date: ...... / ...... / ......

1) What is the gradient of $y = 2 - 4x$?
2) Give the equation of the horizontal line that crosses the y-axis at (0, 6).
3) Give the equation of a straight line with y-intercept = 3 and gradient = 2.
4) Give the coordinates of the x-intercepts of $y = (x + 2)(x - 3)$.
5) Where does the graph $y = 5x$ cross the x-axis?
6) Q lies on the line PR, so that PQ : QR = 1 : 5. True or false? PQ is $\frac{1}{5}$ of the length of QR.
7) Describe the 'table of 3 values' method for drawing straight-line graphs.
8) True or false? Positive x^3 graphs go up from the bottom left.
9) Is the graph of $y = 9$ a horizontal line or a vertical line?
10) What angle is formed by two perpendicular lines?

Total:

Mixed Practice Quizzes

Quiz 3

Date: ...... / ...... / ......

1) Which of these lines is not parallel to the others:
$y = 2x$, $y = 4x + 3$, $y = 2x + 3$?
2) Give the standard format of the equation of a quadratic graph.
3) Give the equation used to work out the gradient from two points.
4) Is the graph of $y = 1 - 3x - x^2$ a straight line, a quadratic or a cubic?
5) Use the '$x = 0$, $y = 0$' method to find where the graph of $y = 4x - 8$ crosses the axes.
6) What does 'm' mean in $y = mx + c$?
7) True or false? A quadratic graph with a positive x^2 coefficient is u-shaped.
8) What is the equation of the main diagonal through the origin?
9) Describe the shape of a cubic graph with a negative coefficient of x^3.
10) A line has gradient $\frac{1}{3}$. What is the gradient of a line perpendicular to it?

Total:

Quiz 4

Date: ...... / ...... / ......

1) Give the equation of the line that is parallel to $y = 6x + 2$ and has a y-intercept of -9.
2) Is the graph of $y = 2x + x^3$ a straight line, a quadratic, or a cubic?
3) $A = (0, 0)$ and $C = (6, 9)$. B lies on AC such that $AB:BC = 2:1$. Find the coordinates of B.
4) Describe the '$x = 0$, $y = 0$' method for drawing straight-line graphs.
5) If the gradients of two perpendicular lines are multiplied together, what is the result?
6) Which letter in $y = mx + c$ gives the y-coordinate of the y-intercept?
7) True or false? The equation of the x-axis is $x = 0$.
8) How can you tell if two straight lines are parallel from their equations?
9) How do you find the equation of a line through two points?
10) The line $y = 4x + 3$ passes through point $(0, p)$. What is the value of p?

Total:

Harder Graphs

First Go: ..... / ..... / .....

Circle Graphs

A circle with centre (,) and r has the equation: .... + ... = ...

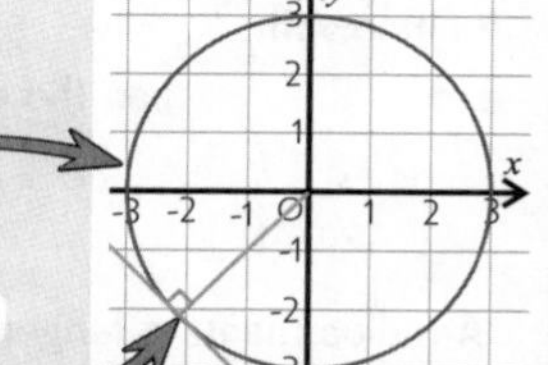

$x^2 + y^2 = 9$ is a circle with centre (,).

$r^2 =$ so radius, r, is .

A radius meets a tangent at , so use to find the equation of a tangent to a circle at a point.

Exponential Graphs

General form: y = ... or y = ... (k is)

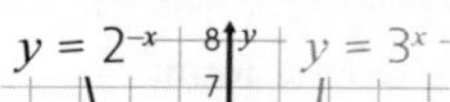

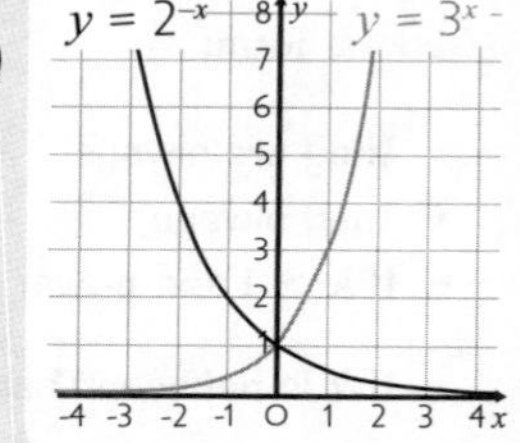

- They are always above the .
- They always go through the point (,).
- If and power is , graph curves upwards.
- If k is between OR power is , then graph is flipped horizontally.

EXAMPLE

The graph shows how the number of bacteria (B) in a sample increases. The equation of the graph is $B = pg^h$, where h = number of hours. p and g are positive constants. Find p and g.

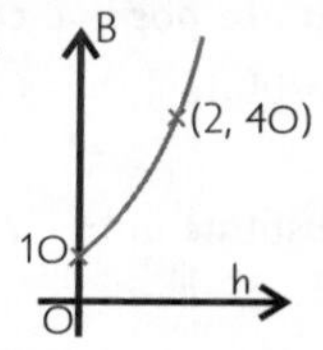

1. Substitute in h = 0, B = 10.
   $10 = pg^0 \Rightarrow 10 =$ .... $\Rightarrow p =$ ....
2. Substitute in h = 2, B = 40.
   .... $= 10g^2 \Rightarrow$ ..... $= g^2 \Rightarrow g =$ .....

Reciprocal Graphs

General form: y = $\frac{\ldots\ldots}{\ldots\ldots}$ or xy = ...

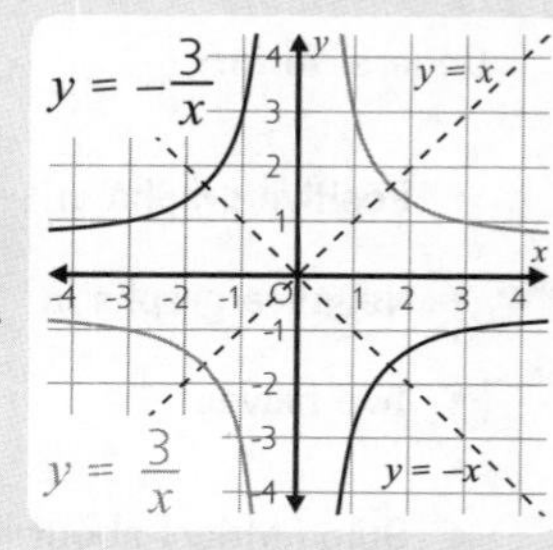

- graphs in top right and bottom left quadrants.
- graphs in top left and bottom right quadrants.
- Two halves of graph .
- Graphs don't exist for .
- about lines y = x and y = –x.

Second Go:
..... /..... /.....

Harder Graphs

Circle Graphs

A circle with
and has the equation:

is a circle with centre .

$r^2 = 9$ so .

A radius meets a tangent

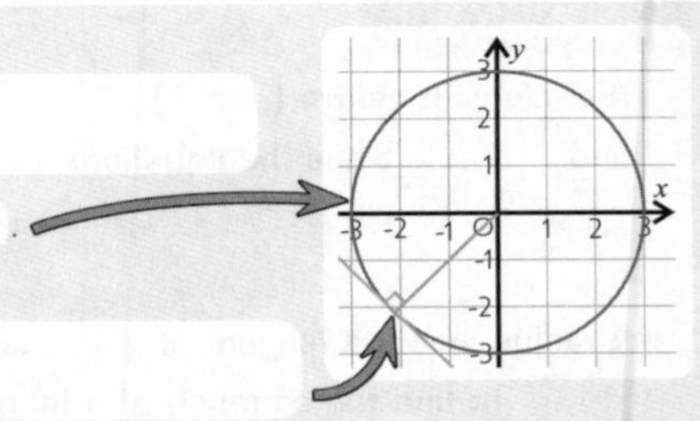

Exponential Graphs

General form: or (k is positive)

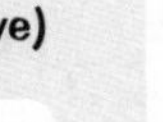

- They are always
- They always
- If k > 1 and power is
- If k is between 0 and 1 OR

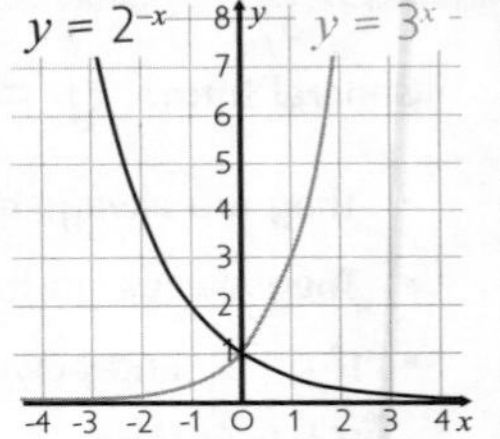

EXAMPLE

The graph shows how the number of bacteria (B) in a sample increases. The equation of the graph is $B = pg^h$, where h = number of hours. p and g are positive constants. Find p and g.

1. Substitute in h = 0, B = 10.
   .... ⇒ .... ⇒ ....
2. Substitute in h = 2, B = 40.
   .... ⇒ .... ⇒ ....

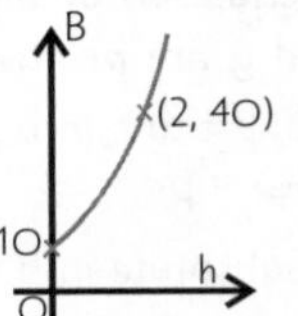

Reciprocal Graphs

General form: or

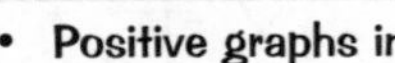

- Positive graphs in
- Negative graphs in
- Two halves
-
- Symmetrical about lines

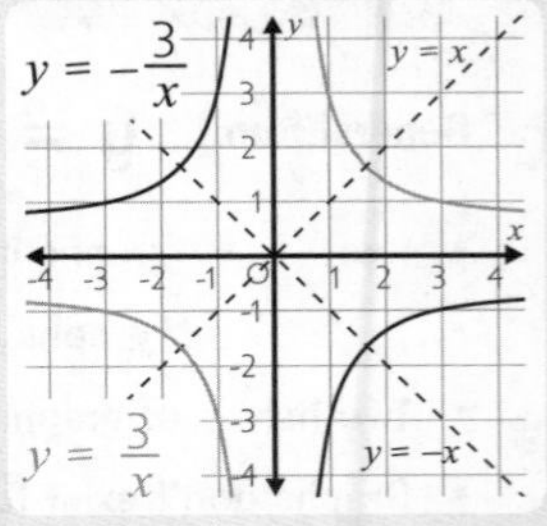

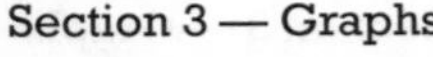

Trig Graphs and Solving Equations

First Go: ..... /..... /.....

Sin x and Cos x Graphs

- Both have y-limits of and .
- every 360°.
- sin graph = cos graph shifted .

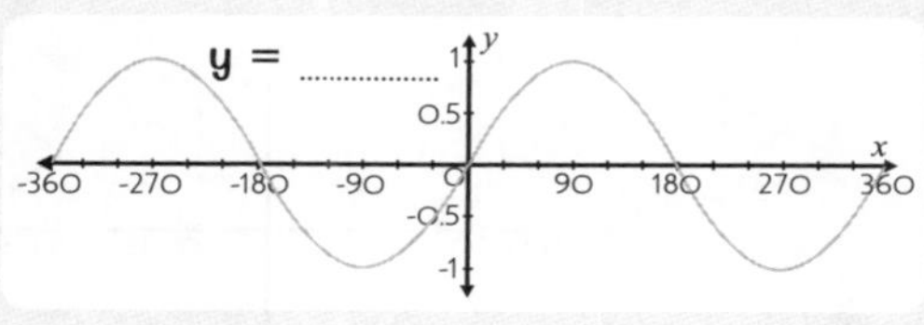

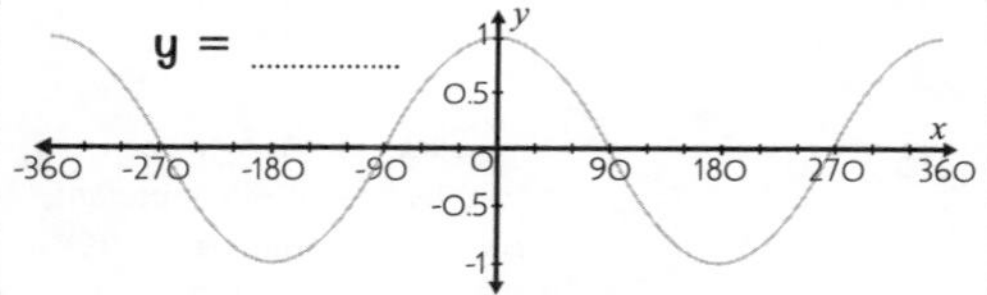

Tan x Graph

- Goes from to .
- Repeats every .
- tan x undefined at , , ...

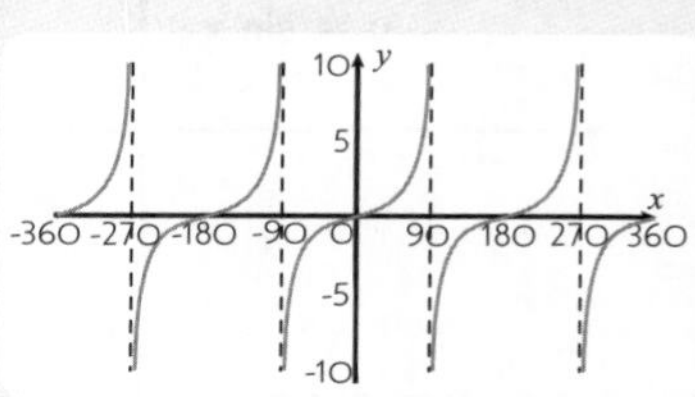

Sketch sin, cos and tan graphs by plotting important points that happen every 90°.

Solve Equations Using Graphs

EXAMPLE

By plotting the graphs, solve the simultaneous equations $y = x^2 - 5$ and $y = x - 3$.

1. Draw both graphs.
2. Find where graphs cross.

$x =$, $y =$ and $x =$, $y =$

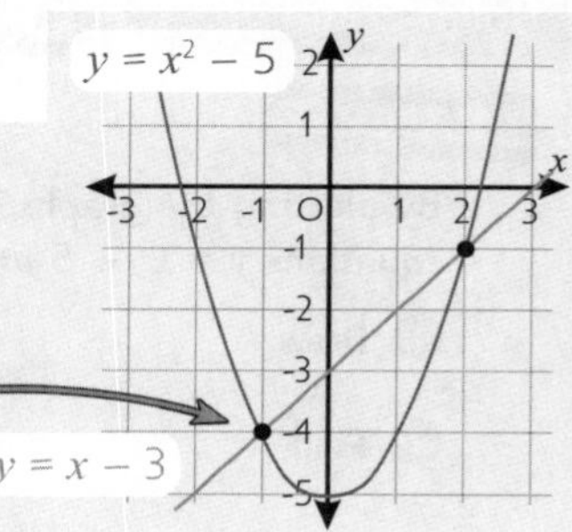

EXAMPLE

The graph of $y = x^3 + 2x$ is shown. Find the equation of the line you'd need to draw to solve $x^3 + 3x - 1 = 0$.

1. the equation you want to solve to get the equation of the graph on .

$x^3 + 3x - 1 = 0$

$x^3 + 2x =$

2. Give the of the line you need to draw.

$y =$

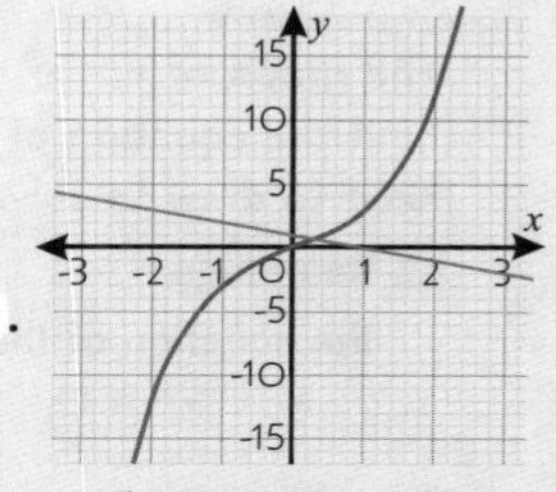

The intersection of these graphs gives the solution to $x^3 + 3x - 1 = 0$.

Second Go:
...../...../.....

Trig Graphs and Solving Equations

Sin x and Cos x Graphs

- Both have .
- Repeat .
- sin graph =
  .

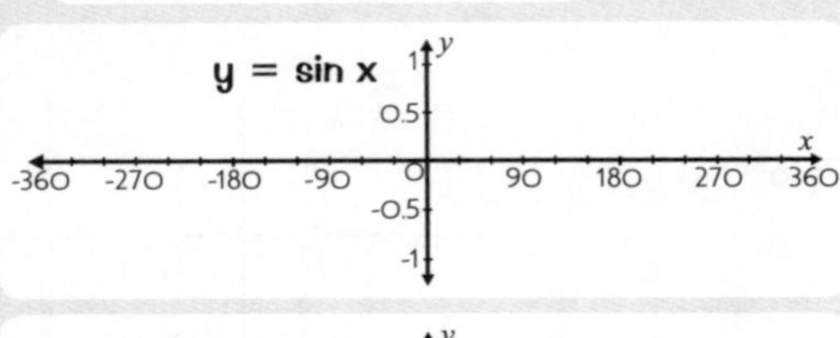

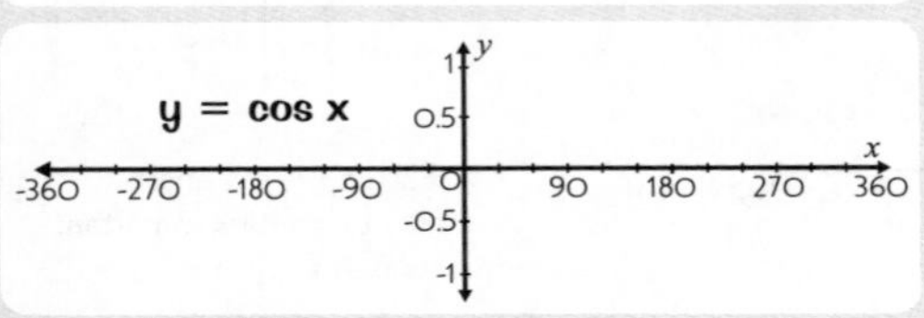

Tan x Graph

- Goes from .
- Repeats .
- tan x
  ...

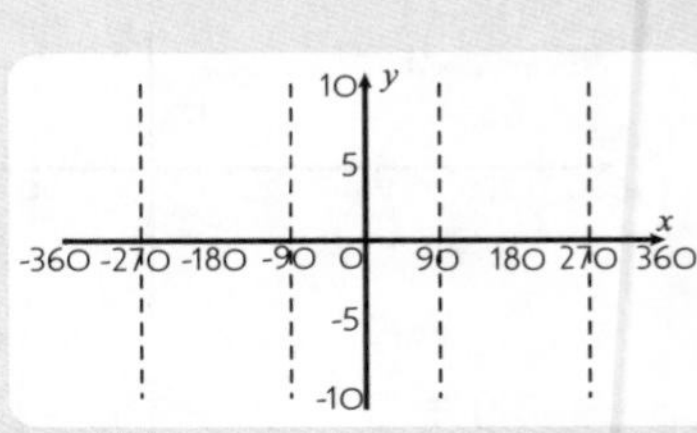

Sketch sin, cos and tan graphs by plotting important points that happen every 90°.

Solve Equations Using Graphs

EXAMPLE

By plotting the graphs, solve the simultaneous equations $y = x^2 - 5$ and $y = x - 3$.

1. Draw .
2. Find .
   and

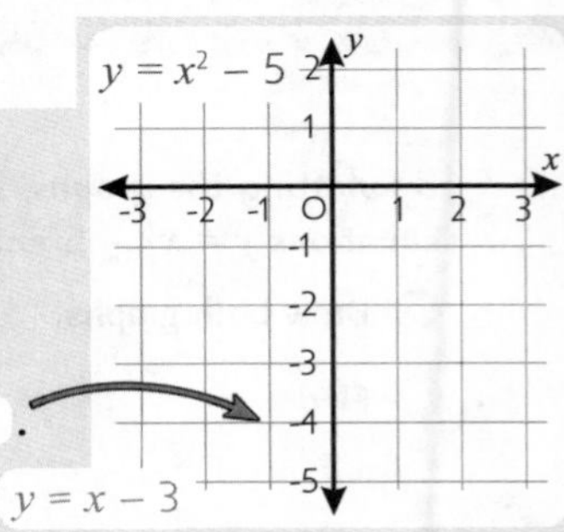

EXAMPLE

The graph of $y = x^3 + 2x$ is shown. Find the equation of the line you'd need to draw to solve $x^3 + 3x - 1 = 0$.

1. Rearrange to get the equation of the graph .
   $x^3 + 3x - 1 = 0$
   =
2. Give the you need to .
   $y =$

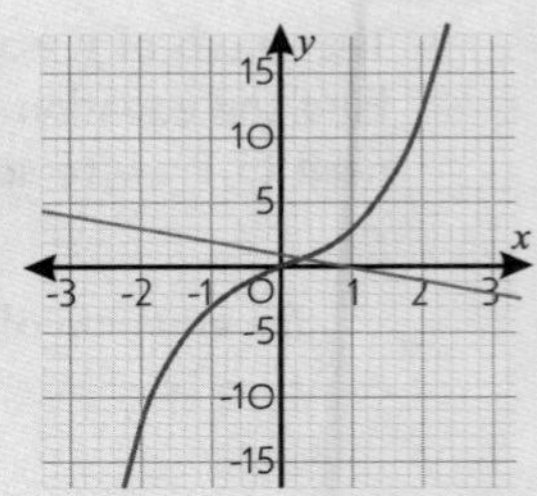

The intersection of these graphs gives the solution to $x^3 + 3x - 1 = 0$.

Graph Transformations

First Go: ...../...../.....

Translations on y-axis: y = f(x) + a

Adding a number to the ______ of the equation translates the graph ______.

For example:

$y = f(x) + 2$ is a translation of 2 units ______.

$y = f(x) - 4$ is a translation of 4 units ______.

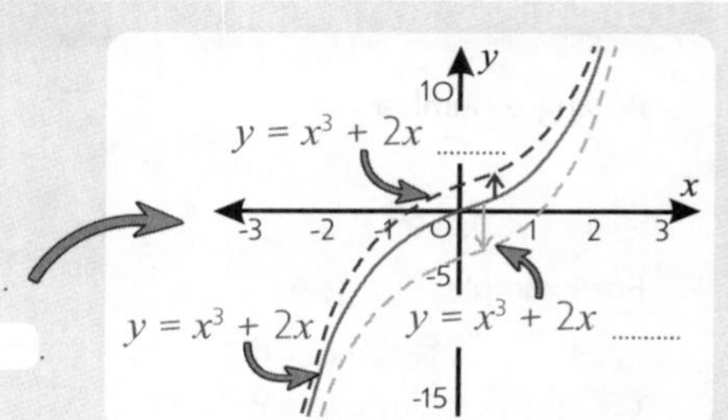

Translations on x-axis: y = f(x – a)

Replacing x ______ in the equation with (x – a) translates the graph ______.

Translations go the 'wrong' way: y = f(x – a) slides y = f(x) 'a' units in the ______ (i.e. ______).

For example:

$y = f(x - 4)$ is a translation of 4 units ______.

$y = f(x + 3)$ is a translation of 3 units ______.

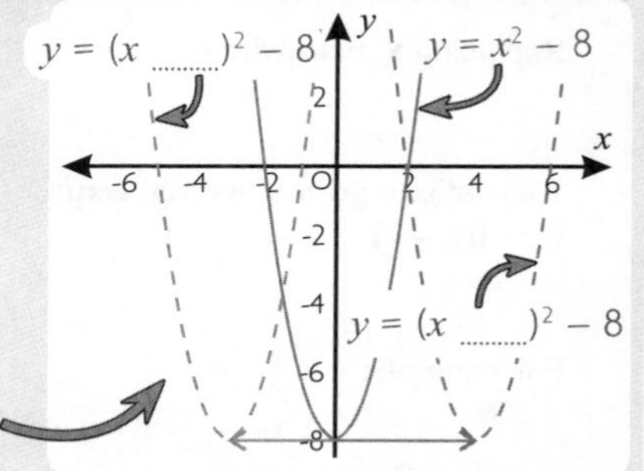

Reflections: y = –f(x)

y = –f(x) is the reflection of y = f(x) in the ______.

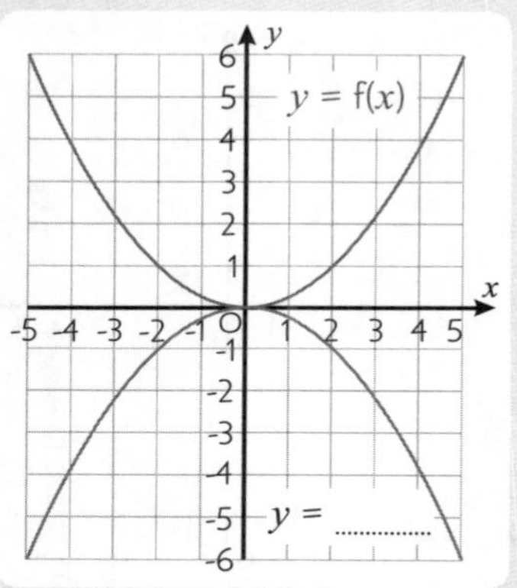

Reflections: y = f(–x)

y = f(–x) is the reflection of y = f(x) in the ______.

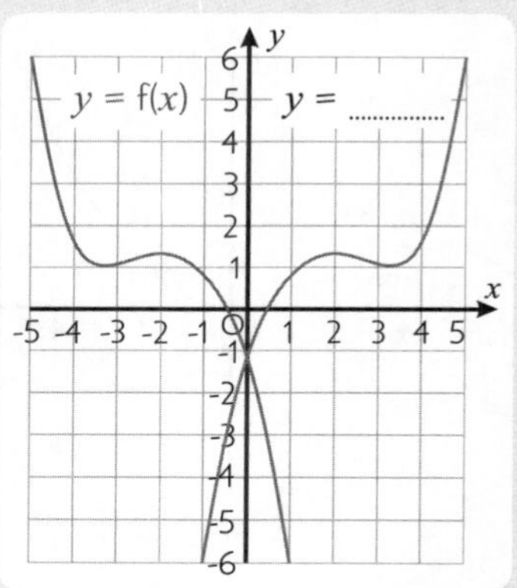

Second Go:
...../...../.....

Graph Transformations

Translations on y-axis: y = f(x) + a

Adding a number

For example:

$y = f(x) + 2$ is a translation of .

$y = f(x) - 4$ is a translation of .

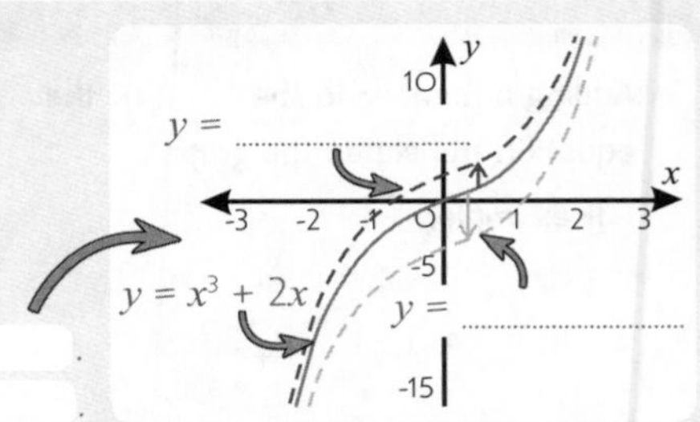

Translations on x-axis: y = f(x – a)

Replacing x everywhere

Translations go the 'wrong' way:
y = f(x – a)

For example:

$y = f(x - 4)$ is a translation of .

$y = f(x + 3)$ is a translation of .

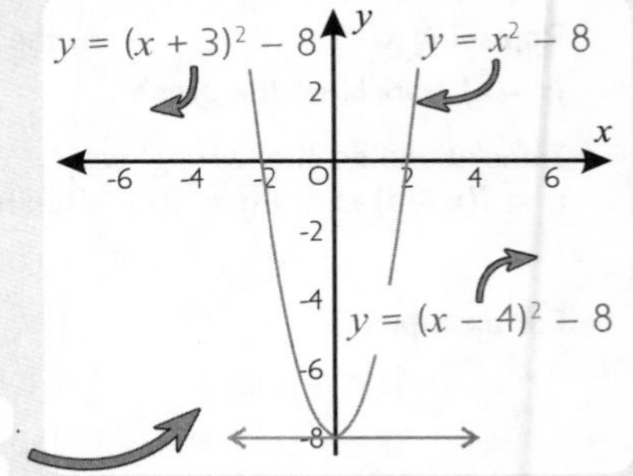

Reflections: y = –f(x)

y = –f(x) is

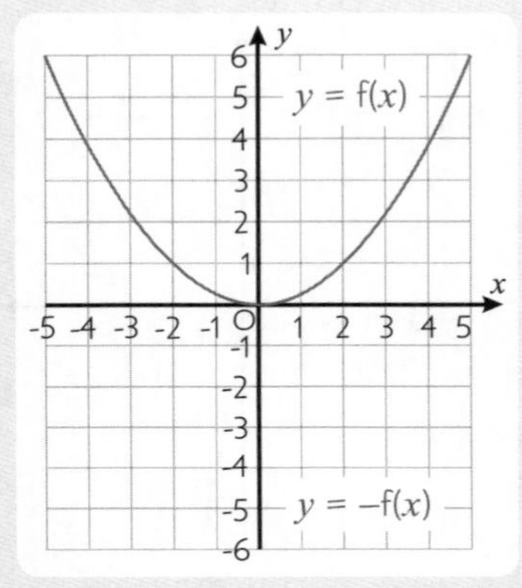

Reflections: y = f(–x)

y = f(–x) is

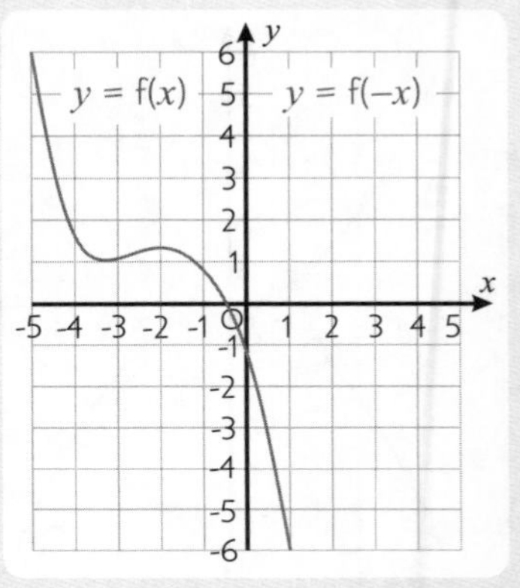

Real-Life Graphs

First Go: ..... /..... /.....

Distance-Time Graphs

Gradient =

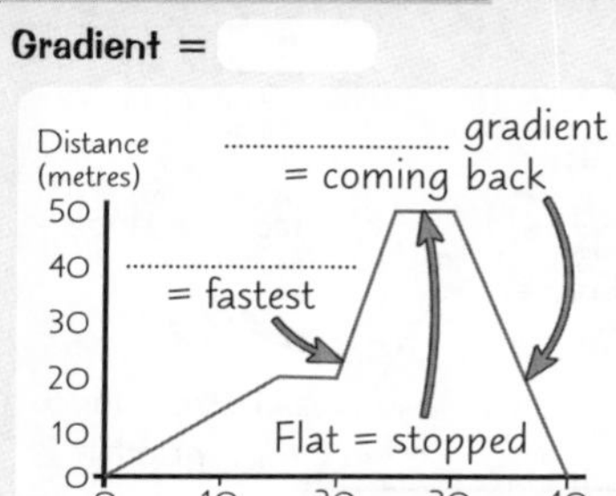

Velocity-Time Graphs

Gradient =

The units of acceleration here are ....

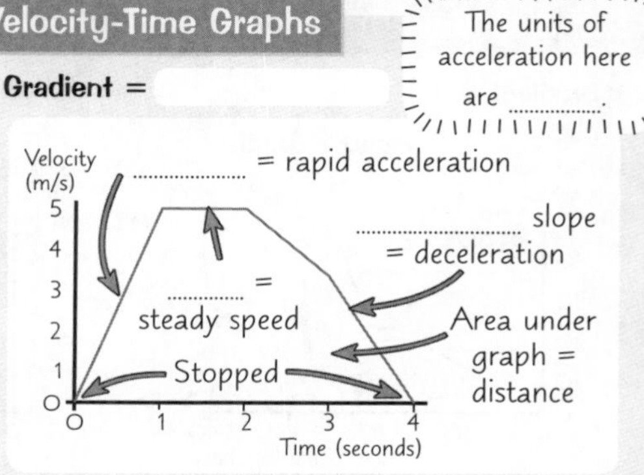

Estimate the Area Under a Velocity-Time Curve

1. **Divide area into ....**
2. **Find .... of each ....**
3. **Add .... together to get the ....**

EXAMPLE

Estimate the distance travelled during the 15 s shown on the graph.

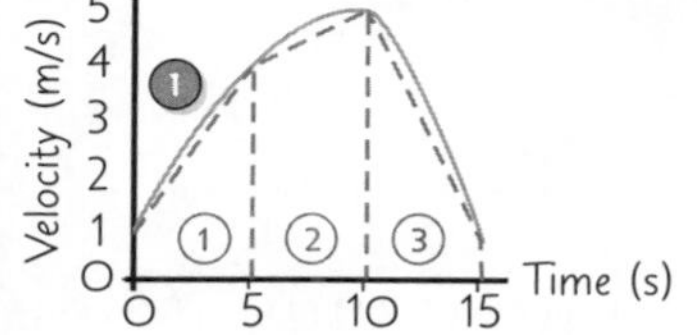

2. Area ① = 0.5 × (.... + ...) × .... = ...
   Area ② = 0.5 × (.... + ...) × .... = ...
   Area ③ = 0.5 × (.... + ...) × .... = ...
3. .... + ... + ... = ... m

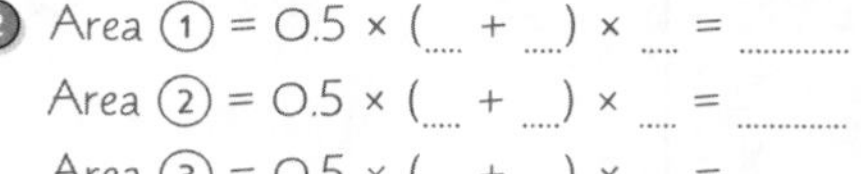

Average velocity = total distance ÷ ....

Gradients of Real-Life Graphs

Gradient represents .... — y-axis unit PER x-axis unit.

E.g. .... PER .... (speed).

Gradient = change in ...... / change in ......

Finding an Average Gradient

E.g. average speed between 2 s and 4 s:

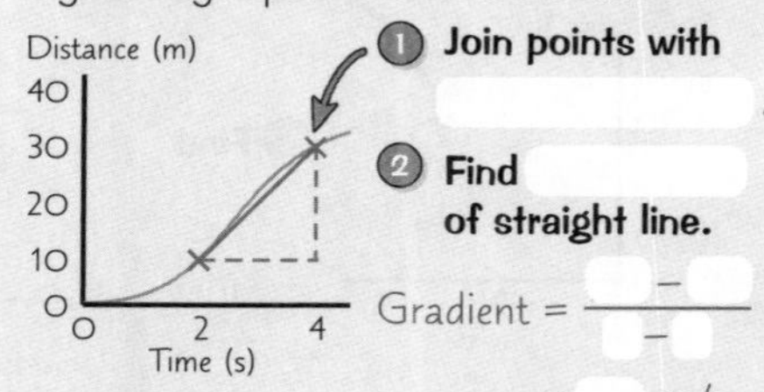

1. **Join points with ....**
2. **Find .... of straight line.**

Gradient = (.... – ....) / (.... – ....)

= .... m/s

Estimating a Gradient

E.g. speed at 3 s:

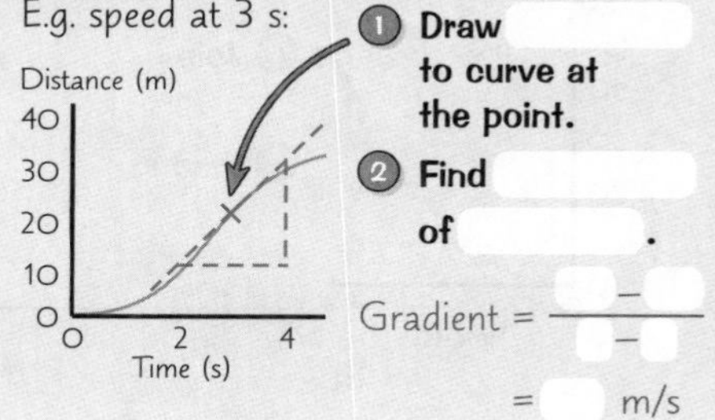

1. **Draw .... to curve at the point.**
2. **Find .... of ....**

Gradient = (.... – ....) / (.... – ....)

= .... m/s

Second Go:
..... /..... /.....

Real-Life Graphs

Distance-Time Graphs

Gradient =

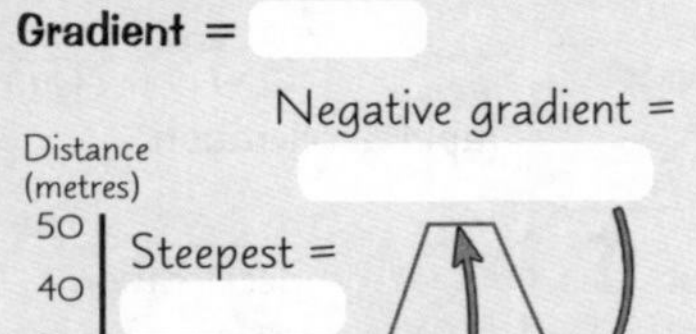

Velocity-Time Graphs

Gradient =

The units of .... here are ....

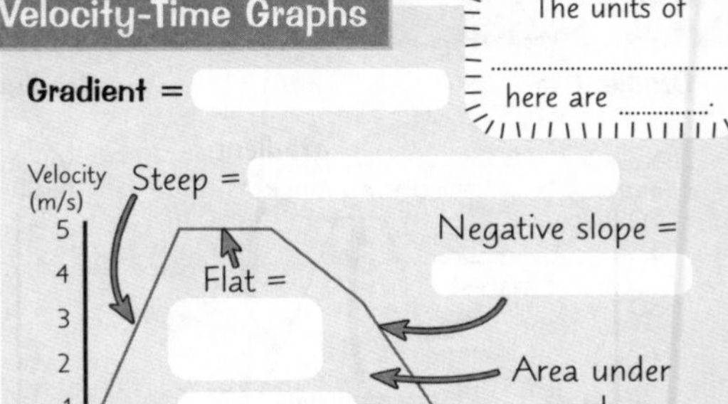

Estimate the Area Under a Velocity-Time Curve

1. **Divide**
2. **Find**
3. **Add**

EXAMPLE **Estimate the distance travelled during the 15 s shown on the graph.**

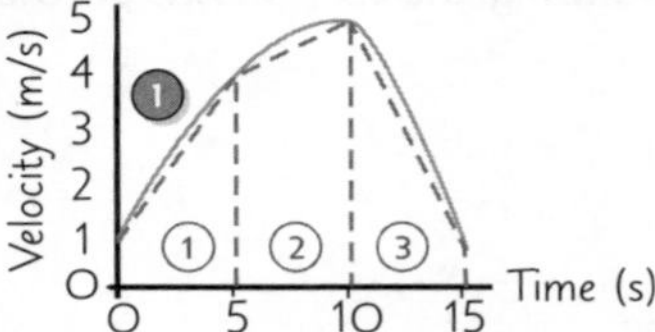

2. Area ① = ... = ...
   Area ② = ... = ...
   Area ③ = ... = ...
3. .... = ...

Average velocity = ... ÷ ....

Gradients of Real-Life Graphs

Gradient represents rate —

Gradient =

Finding an Average Gradient

E.g. average speed between 2 s and 4 s:

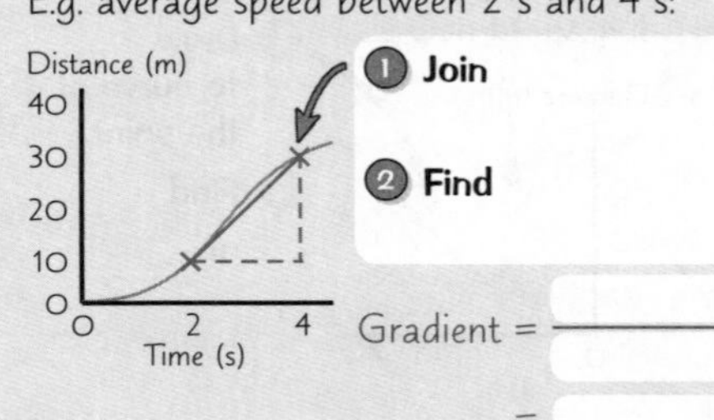

1. **Join**
2. **Find**

Gradient = ——

=

Estimating a Gradient

E.g. speed at 3 s:

1. **Draw**
2. **Find**

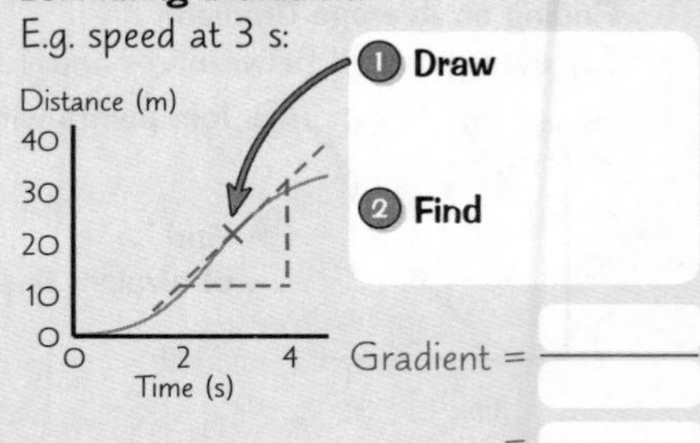

Gradient = ——

=

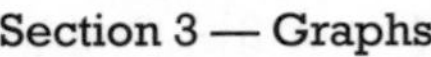

Mixed Practice Quizzes

You've come the distance at great velocity, so it's time for more quizzes.
These ones are for p.61-68. Add up your marks once you're finished.

Quiz 1

Date: ...... / ...... / ......

1) What is the radius of the circle given by the equation $x^2 + y^2 = 4$?
2) Describe the translation that maps $y = x^2 + 2x - 3$ to $y = x^2 + 2x + 5$.
3) For the graph of $y = \sin x$, what is the value of y when $x = 90°$?
4) What does the gradient of a distance-time graph represent?
5) True or false? An exponential graph is always above the x-axis.
6) How do you find the average gradient between two points on a curved graph?
7) The curve $y = x^2 - 3x + 2$ is shown on a graph. What line would you draw on the graph to solve the quadratic $x^2 + x = 0$?
8) What is the first step to estimate the area under a velocity-time graph?
9) Is $y = f(-x)$ a reflection of $f(x)$ in the x-axis or in the y-axis?
10) True or false? The graph of $\tan x$ is undefined at 0°.

Total:

Quiz 2

Date: ...... / ...... / ......

1) A reciprocal graph has two lines of symmetry. Give the equations of these lines.
2) Give the equation of the reflection of $y = 8x - 2$ in the x-axis.
3) How do you use graphs to solve simultaneous equations?
4) Describe the transformation from the graph of $f(x)$ to the graph of $f(x + 2)$.
5) Describe how to estimate the gradient at a point on a curved graph.
6) What does a flat section on a distance-time graph represent?
7) Give the coordinates of the point that all exponential graphs go through.
8) True or false? $\sin x$ and $\cos x$ graphs repeat every 180°.
9) What does a slope with a negative gradient mean on a velocity-time graph?
10) What are the y-limits of $\sin x$ and $\cos x$ graphs?

Total:

Mixed Practice Quizzes

Quiz 3

Date: ...... / ...... / ......

1) Give the general equation for an exponential graph.
2) Which of these translations of f(x) translates the graph vertically: f(x) + a or f(x + a)?
3) True or false? A flat section on a velocity-time graph shows that an object has stopped moving.
4) For exponential graphs where $k > 1$ and the power is positive, which way does the graph curve?
5) After how many degrees does the graph of tan x repeat?
6) In which two quadrants is the graph of $xy = 5$ found?
7) What is the equation of a circle with centre (0, 0) and radius 10?
8) Give the equation of the new graph when $y = x^2$ is translated 5 units left.
9) True or false? The cos x graph shifted 90° right gives the sin x graph.
10) On a graph showing distance travelled in metres against time in seconds, what are the units of the gradient?

Total:

Quiz 4

Date: ...... / ...... / ......

1) How does the graph of $y = f(x) - 3$ differ from the graph of $y = f(x)$?
2) What does a slope with a negative gradient mean on a distance-time graph?
3) For what value of x do reciprocal graphs not exist?
4) What does the area under a velocity-time curve represent?
5) Describe the difference between an exponential graph with $0 < k < 1$ and an exponential graph with $k > 1$. (The power is positive for both graphs.)
6) Which of the following graphs don't go through (0, 0): $y = \sin x$, $y = \cos x$, $y = \tan x$?
7) What would you do to an equation to translate its graph 2 units up?
8) True or false? The two halves of a reciprocal graph never touch.
9) Which trig graph has y-values from $-\infty$ to $+\infty$?
10) What type of graph is given by the equation $xy = 3$?

Total:

Ratios

First Go: ..... /..... /.....

Writing Ratios as Fractions

Write one number .... of the other.

Or .... the parts to find a fraction of the ....

EXAMPLE

In a car park, the ratio of cars to vans is 8:3.

- There are as many cars as vans.
- There are as many vans as cars.
- There are + = parts in total, so are cars and are vans.

Four Ways to Simplify Ratios

1 all numbers by the same thing.

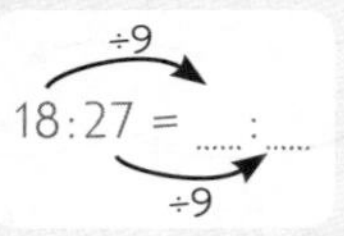

The .... button on your calculator can be used to help simplify ratios.

2 Multiply to get rid of and .

1.5:3.5 = ...:........ = ...:......
×10 ×10 ÷5 ÷5

3 Convert to the .

0.75 kg:250 g = ... g:............ g
÷250 ÷250
= .....:.....

4 Divide to get in the form or .

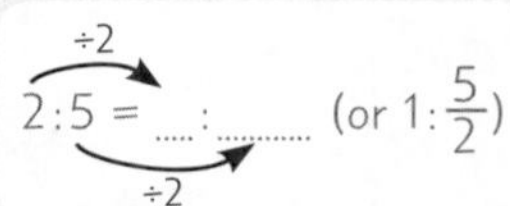

Three Steps to Scale Up Ratios

1 Work out what one side of the ratio is to get its actual value.

2 Multiply the by this number.

3 the two sides to find the (if the question asks you to).

EXAMPLE

A theatre audience is made up of adults and children in the ratio 3:5. There are 105 adults. How many people are there in the audience in total?

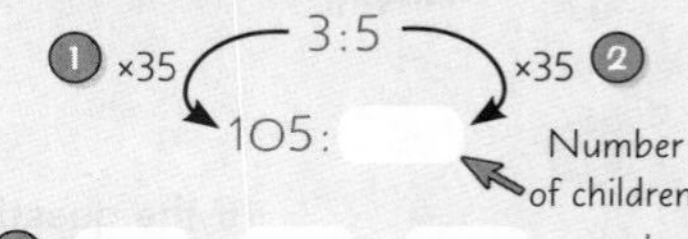

3 + = people

The two sides of a ratio are always in ....

Second Go:
...../...../.....

Ratios

Writing Ratios as Fractions

Write ....

Or .... to find a ....

EXAMPLE

In a car park, the ratio of cars to vans is 8:3.

- There are as many .
- There are as many .
- There are in total, so are cars and are vans.

Four Ways to Simplify Ratios

1. all numbers by the .

   18:27 = ... (÷...... , ÷......)

2. to get rid of and .

   1.5:3.5 = ... = ... (×........ , ÷...... ; ×........ , ÷......)

The .... button on your calculator can be used to help ....

3. to the .

   0.75 kg:250 g = ... = ... (÷........... , ÷...........)

4. to get in the form or .

   2:5 = ... (or ....) (÷...... , ÷......)

Three Steps to Scale Up Ratios

1. Work out what one side of the ratio is

2. Multiply

3. (if the question asks you to).

EXAMPLE

A theatre audience is made up of adults and children in the ratio 3:5. There are 105 adults. How many people are there in the audience in total?

1 × 3:5 × 2

Number of children.

3

The .... are always in ....

More Ratios and Proportion

First Go: ..... /..... /.....

Part : Whole Ratios

PART : WHOLE RATIO — of ratio included in .

EXAMPLE

part : part → $\frac{\text{part}}{\text{whole}}$ → part : whole

3 : 7 → .... / .... = ... / .... → .... : ....

Three Steps for Proportional Division

1. the parts.
2. to find one part.
3. to find the amounts.

EXAMPLE

1200 g of flour is used to make cakes, pastry and bread in the ratio 8 : 7 : 9. How much flour is used to make pastry?

1. + + = parts
2. 1 part = g ÷ = g
3. 7 parts = × g = g

Two Steps for Direct Proportion

DIRECT PROPORTION — one quantity the other proportionally.

1. to find the amount for one thing.
2. to find the amount for the number of things you want.

EXAMPLE

Vivek uses 1125 ml of milk to make 5 milkshakes. How much milk will he need to make 12 milkshakes?

1. 1 milkshake uses ml ÷ = ml
2. 12 milkshakes will use ml × = ml

Two Steps for Inverse Proportion

INVERSE PROPORTION — one quantity the other proportionally.

1. to find the amount for one thing.
2. to find the amount for the number of things you want.

EXAMPLE

Three farmers can shear 75 sheep in 45 minutes. How long would it take five farmers to shear the same number of sheep?

1. 75 sheep would take 1 farmer × = minutes
2. 5 farmers would take ÷ = minutes

Second Go:
..... /..... /.....

More Ratios and Proportion

Part : Whole Ratios

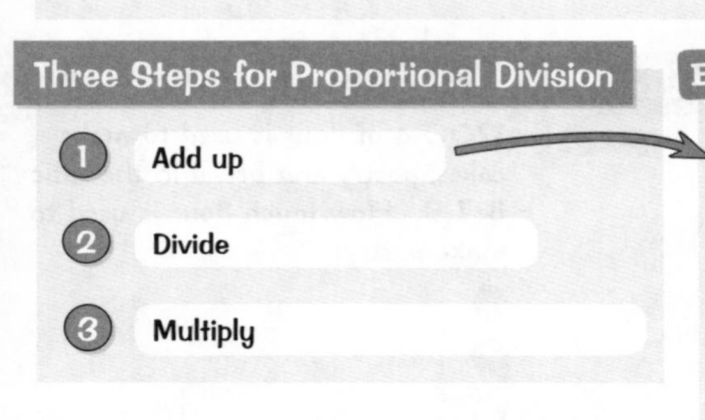

PART : WHOLE RATIO —

EXAMPLE

part:part — $\frac{\text{part}}{\text{whole}}$ — part:whole

3:7 → .... → ....

Three Steps for Proportional Division

1. Add up
2. Divide
3. Multiply

EXAMPLE

1200 g of flour is used to make cakes, pastry and bread in the ratio 8:7:9. How much flour is used to make pastry?

1.
2. 1 part =
3. 7 parts =

Two Steps for Direct Proportion

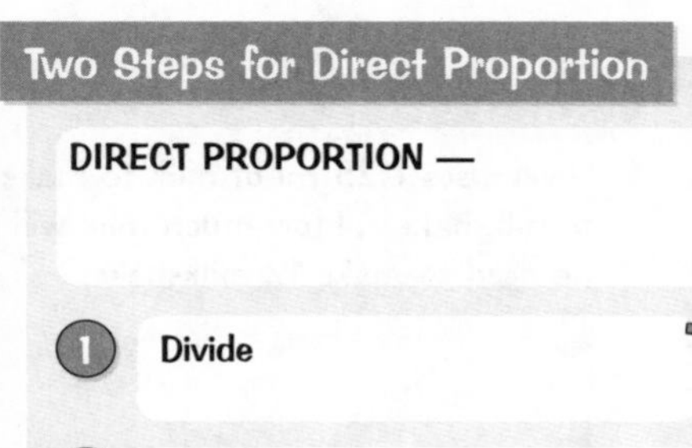

DIRECT PROPORTION —

1. Divide
2. Multiply

EXAMPLE

Vivek uses 1125 ml of milk to make 5 milkshakes. How much milk will he need to make 12 milkshakes?

1. 1 milkshake uses
2. 12 milkshakes will use

Two Steps for Inverse Proportion

INVERSE PROPORTION —

1. Multiply
2. Divide

EXAMPLE

Three farmers can shear 75 sheep in 45 minutes. How long would it take five farmers to shear the same number of sheep?

1. 75 sheep would take 1 farmer
2. 5 farmers would take

Direct and Inverse Proportion

First Go: ..... /..... /.....

Turning Proportions into Equations

	Proportionality	Equation
y is to x	$y \propto x$	$y =$
y is to x	$y \propto$	$y = \frac{k}{x}$
y is proportional to the	$y \propto x^2$	$y =$
y is inversely proportional to x cubed	$y \propto$	$y =$

k is a constant.

$\propto$ means 'is proportional to'.

Drawing Proportion Graphs

y is

to x

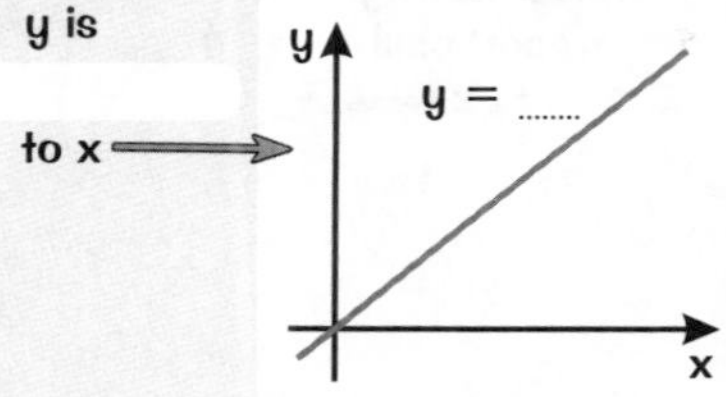

y is

to x

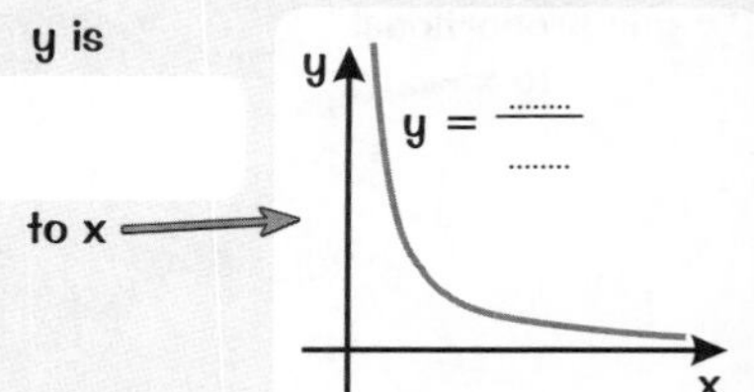

y is

to x^2

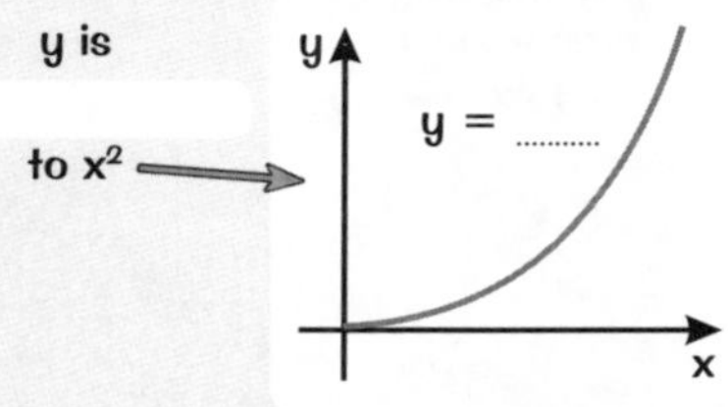

y is

to x^3

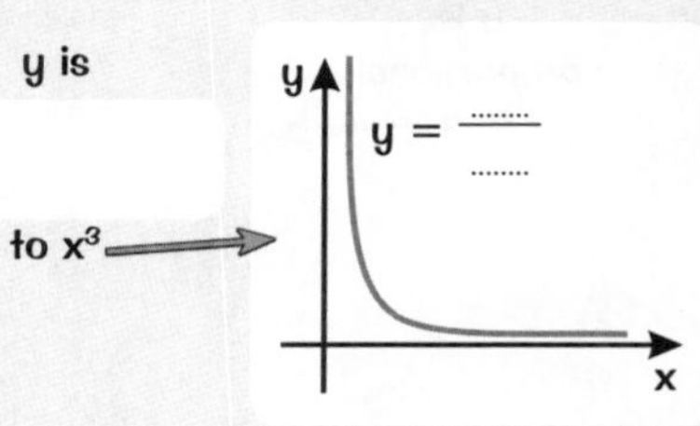

Four Steps for Algebraic Proportion

1. Convert
   to .
2. Use given values to find .
3. Put back into .
4. Use to find .

EXAMPLE

P is proportional to the square of Q.
When P = 320, Q = 4.
Find P when Q = 10.

1. $P \propto Q^2$, so $P =$
2. $= k(\quad) = \quad k$, so $k =$
3. $P =$
4. $P = \quad = \quad =$

Second Go:
...../...../.....

Direct and Inverse Proportion

Turning Proportions into Equations

	Proportionality	Equation
		$y = kx$
y is inversely proportional to x		
	$y \propto x^2$	
	$y \propto \frac{1}{x^3}$	

k is a constant.

$\propto$ means 'is proportional to'.

Drawing Proportion Graphs

y is proportional to x

y is inversely proportional to x

y is proportional to x^2

y is inversely proportional to x^3

Four Steps for Algebraic Proportion

1. Convert
2. Use
3. Put
4. Use

EXAMPLE

P is proportional to the square of Q.
When P = 320, Q = 4.
Find P when Q = 10.

1. , so
2. , so
3. P =
4. P =

Percentages

First Go: ..... /..... /.....

Three Simple Percentage Questions

1. To find a percentage of an amount, turn the percentage into a __________ / __________ then __________.

35% of 240 = ... × ....
= ....

2. To find the amount after a percentage change, find the __________ and multiply the __________ by it.

EXAMPLE

Items in a sale have 12% off. What is the sale price of a hat that usually costs £7.50?

Multiplier for 12% decrease = ... − .... = ...

Sale price of hat = £.......... × .... = £..........

% increase = multiplier .... than 1

% decrease = multiplier .... than 1

3. To write one number as a percentage of another, __________ the first by the second then __________.

30 as a % of 250 = $\frac{\dots}{\dots}$ × 100 = ...%

Two Steps to Find the Percentage Change

1. Find the change in __________.
2. Use this formula:
Percentage change = $\frac{\quad}{\quad}$ × 100

'Change' = ..., decrease, ..., loss, etc.

EXAMPLE

A car was bought for £11 500. Four years later, it is sold for £8855. Find the percentage loss.

1. loss = £ ______ − £ ______
= £ ______

2. % loss = $\frac{\quad}{\quad}$ × 100
= ______ × 100 = ______%

Three Steps to Find the Original Value

1. Write the amount as a __________ of the __________.
2. Divide to find ______ of the original value.
3. Multiply by ______ to find the original value (100%).

EXAMPLE

A village has a population of 1003. The population of the village has increased by 18% since 2016. What was the population in 2016?

1. 1003 = ______%
2. 1003 ÷ ______ = ______% ÷ ______
______ = 1%
3. ______ × ______ = 1% × ______
______ = 100%

Second Go:
...../...../.....

Percentages

Three Simple Percentage Questions

1. To find a percentage of an amount,

35% of 240 = ...
= ....

2. To find the amount after a percentage change,

EXAMPLE

Items in a sale have 12% off. What is the sale price of a hat that usually costs £7.50?

Multiplier for 12% decrease = ...

Sale price of hat = ...

% increase = ...
....

% decrease = ...
....

3. To write one number as a percentage of another,

30 as a % of 250 = ...
= ....

Two Steps to Find the Percentage Change

1. Find

2. Use this formula:

Percentage change =

'Change' = ...
.... etc.

EXAMPLE

A car was bought for £11 500. Four years later, it is sold for £8855. Find the percentage loss.

1. loss =
=

2. % loss =
=

Three Steps to Find the Original Value

1. Write

2. Divide

3. Multiply

EXAMPLE

A village has a population of 1003. The population of the village has increased by 18% since 2016. What was the population in 2016?

1. 1003 =
2. =
=
3. =
= 100%

Working with Percentages

First Go: ..... /..... /.....

Simple Interest

SIMPLE INTEREST — a % of the ______ is paid at regular intervals (e.g. every year). The amount of interest ______.

Three steps for simple interest questions:

1. Find the % of the ______.
2. ______ by the number of intervals.
3. ______ to original value (if needed).

EXAMPLE

Lila puts £2500 in a savings account that pays 3.5% simple interest each year. How much will be in the account after 5 years?

1. 3.5% of £2500 = ______ × £______ = £______
2. 5 × £______ = £______
3. £2500 + £______ = £______

Compound Growth and Decay

$N = \_\_ \times (\text{multiplier})^{\_}$ ← Number of years/days/hours etc.

Initial amount

EXAMPLE

A boat was bought for £15 000. It depreciates in value by 11% each year. How much will it be worth after 6 years?

N_0 = £15 000, multiplier = ... − .... = ..., n = 6

Value after 6 years = £.................. × ....$^{....}$ = £.................... (to the nearest penny)

Compound Interest

COMPOUND INTEREST — a % of the ______ is paid at regular intervals (e.g. every year). The amount of interest ______.

It's an example of ______.

EXAMPLE

Beth invests £4800 in a savings account that pays 2% compound interest per annum. How much will there be in the account after 3 years?

N_0 = £4800, multiplier = ... + ... = ..., n = 3

Amount after 3 years = £.................. × ....$^{....}$ = £.................... (to the nearest penny)

Second Go:
...../...../.....

Working with Percentages

Simple Interest

SIMPLE INTEREST —

Three steps for simple interest questions:

1. **Find**
2. **Multiply**
3. **Add** **(if needed).**

EXAMPLE

Lila puts £2500 in a savings account that pays 3.5% simple interest each year. How much will be in the account after 5 years?

1. 3.5% of £2500
   =
   =
2.
3.

Compound Growth and Decay

Amount after n years/days/hours etc. → ← Number of years/days/hours etc.

Initial amount % change multiplier

EXAMPLE

A boat was bought for £15 000. It depreciates in value by 11% each year. How much will it be worth after 6 years?

N_0 = ..., multiplier = ..., n = ...

Value after 6 years = ... (to the nearest penny)

Compound Interest

COMPOUND INTEREST —

EXAMPLE

Beth invests £4800 in a savings account that pays 2% compound interest per annum. How much will there be in the account after 3 years?

N_0 = ..., multiplier = ..., n = ...

Amount after 3 years = ... (to the nearest penny)

Measures and Units

First Go: ..... /..... /.....

Unit Conversions

To convert between units, multiply/divide by a ....

Metric unit conversions:

1 cm = ___ mm
1 tonne = ___ kg
1 ___ = 100 cm
1 ___ = 1000 ml
1 km = ___ m
1 litre = ___ cm^3
1 ___ = 1000 g
1 cm^3 = ___ ml

For metric-imperial conversions, conversion factors will be given.

EXAMPLE

Use the conversion 5 miles ≈ 8 km to work out how many metres there are in 13 miles.

To convert from miles to km, divide by 5 then multiply by 8:

13 miles ≈ ___ ÷ ___ × ___ = ___ × ___

= ___ km

Then convert km to m using the conversion factor 1000:

20.8 km = ___ × ___

= ___ m

Converting Areas

1 m^2 = ___ cm × ___ cm

= ___ cm^2

1 cm^2 = ___ mm × ___ mm

= ___ mm^2

Converting Volumes

1 m^3 = ___ cm × ___ cm × ___ cm

= ___ cm^3

1 cm^3 = ___ mm × ___ mm × ___ mm

= ___ mm^3

Speed, Density and Pressure

Use formula triangles to ....
Cover up the thing you want and write down what's left.

SPEED = ———

Units of speed: distance travelled per unit time, e.g. ___ , ___

DENSITY = ———

Units of density: mass per unit volume, e.g. ___ , ___

EXAMPLE

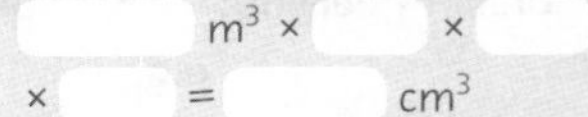

The density of copper is 8.96 g/cm^3. What is the mass of a copper cube with volume 0.008 m^3?

Convert volume to cm^3:

___ m^3 × ___ × ___

× ___ = ___ cm^3

Use the formula triangle to get the formula for mass:

mass = ___ × ___

= ___ × ___

= ___ g

PRESSURE = ———

Units of pressure: force per unit area, e.g. ___ (or ___)

Second Go: ...... /...... /......

Measures and Units

Unit Conversions

To convert between units, ... by a ....

Metric unit conversions:

1 cm = 	1 tonne =

1 m = 	1 litre =

1 km = 	1 litre =

1 kg = 	1 cm³ =

For .... conversions, conversion factors will be given.

EXAMPLE

Use the conversion 5 miles ≈ 8 km to work out how many metres there are in 13 miles.

To convert from miles to km, divide by 5 then multiply by 8:

13 miles ≈ =

=

Then convert km to m using the conversion factor 1000:

20.8 km =

=

Converting Areas

$1\ m^2$ =

= cm^2

$1\ cm^2$ =

= mm^2

Converting Volumes

$1\ m^3$ =

= cm^3

$1\ cm^3$ =

= mm^3

Speed, Density and Pressure

SPEED = ——————

Units of speed:

per , e.g. ,

DENSITY = ——————

Units of density: per

, e.g. ,

PRESSURE = ——————

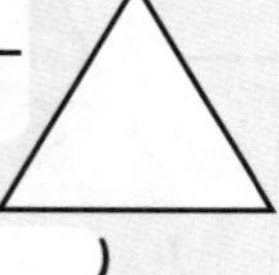

Units of pressure: per

, e.g. (or)

Use formula triangles to ....
Cover up ....
and write down ....

EXAMPLE

The density of copper is 8.96 g/cm³. What is the mass of a copper cube with volume 0.008 m³?

Convert to :

Use the formula triangle to get the formula for mass:

mass =

=

=

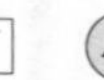

Mixed Practice Quizzes

Get ready for some quick-fire quizzes to test you on pages 71-82.
Answer the questions, then mark each test yourself and add up your score.

Quiz 1

Date: ...... / ...... / ......

1) How do you turn a ratio into a fraction?
2) What is the formula for percentage change?
3) If two quantities are in direct proportion and one of them is tripled, what happens to the other quantity?
4) After adding up the parts, what is the next step in proportional division?
5) How do you convert from m^2 to cm^2?
6) What is the multiplier for a 7.5% increase?
7) If a ratio uses two different units, how do you simplify it?
8) What is the difference between simple interest and compound interest?
9) True or false? Parts in a ratio are always in inverse proportion.
10) A house increases in value. Given that £230 000 = 115%, what is the next thing you would have to do to find the original value of the house?

Total:

Quiz 2

Date: ...... / ...... / ......

1) What does a direct proportion graph look like?
2) How do you find a percentage of an amount?
3) How do you find a fraction of the total using ratios?
4) What is the first thing you have to do to find the original value after a percentage change?
5) In a part : whole ratio, how do you find the size of the other part?
6) What is the formula for compound growth and decay?
7) How do you simplify a ratio that contains fractions?
8) Write 'y is inversely proportional to the square of x' as a proportionality statement.
9) How many cm^3 are there in 1 litre?
10) What is the formula for speed?

Total:

Mixed Practice Quizzes

Quiz 3

Date: ...... / ...... / ......

1) If two quantities are in inverse proportion and one of them is halved, what happens to the other quantity?
2) What would be the first step in simplifying the ratio 300 ml : 0.8 litres?
3) How do you find the new amount after a percentage change?
4) How would you find the interest earned from an account paying 2.5% simple interest over 5 years?
5) How many kg are there in 1 tonne?
6) How do you scale up a ratio if you're given the actual value of one side?
7) What is the formula for density?
8) What is the first step in proportional division?
9) A savings account is opened with a starting value of £1000. It pays 1.5% compound interest per year for 4 years. What are the values of:
   a) N_0? b) the multiplier? c) n?
10) Write 'y is proportional to the cube root of x' as an equation.

Total:

Quiz 4

Date: ...... / ...... / ......

1) How do you write one number as a percentage of another?
2) What is the formula for pressure?
3) What does an inverse proportion graph look like?
4) A proportion has been converted to the equation $y = kx^3$. When $x = 3$, $y = 108$. What is the value of k?
5) What would be the first step in simplifying the ratio 2.8 : 4.2?
6) What is the multiplier for a 14% decrease?
7) What is the total number of parts in the ratio 3 : 10 : 4?
8) In the compound growth and decay formula, what does N_0 represent?
9) How do you convert from mm^3 to cm^3?
10) 800 ml of lemonade is shared in the ratio 3 : 7. How many ml is one part?

Total:

Geometry

First Go:
..... /..... /.....

Five Angle Rules

1. Angles in a triangle add up to °.
2. Angles on a add up to 180°.
3. Angles in a add up to 360°.
4. Angles round a point add up to °.
5. Isosceles triangles have 2 and 2 .

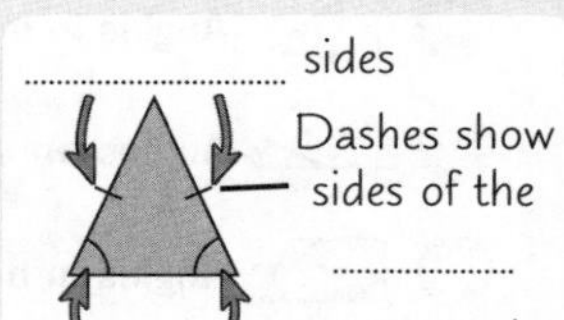

Angles Around Parallel Lines

When a line crosses two :

- Two bunches of are formed.
- There are only different angles (a and b).
- angles are equal.

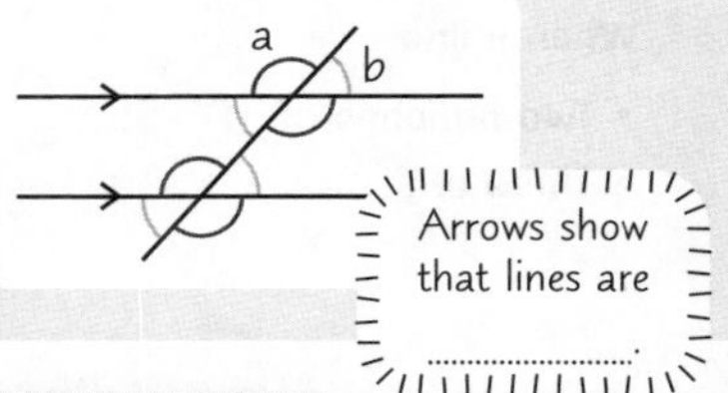

Alternate Angles

Found in a -shape:

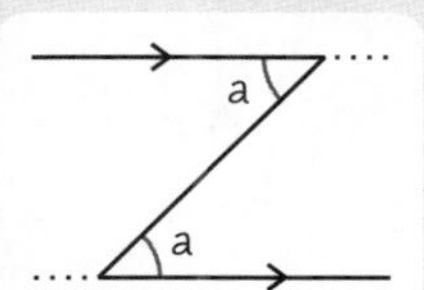

Alternate angles are the .

Corresponding Angles

Found in an -shape:

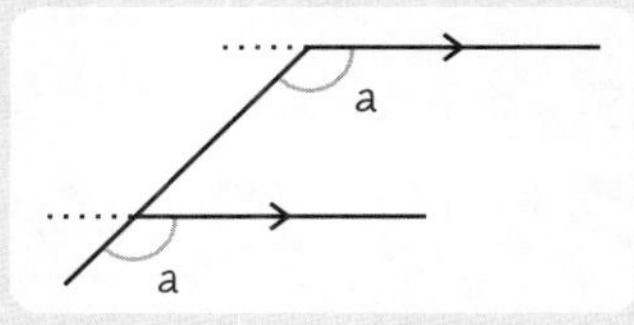

Corresponding angles are the .

Allied Angles

Found in a - or -shape:

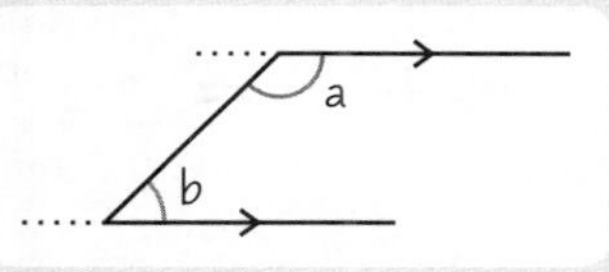

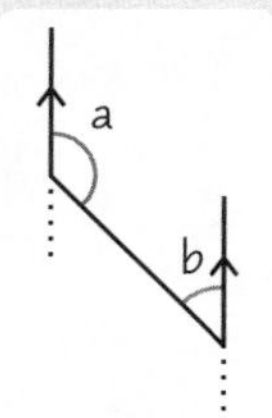

Allied angles add up to °.

$a + b =$ .... °

Second Go:
..... /..... /.....

Geometry

Five Angle Rules

1  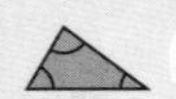 Angles in a .

2   Angles on a .

3   Angles in a .

4  Angles round a .

5 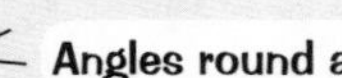 Isosceles triangles .

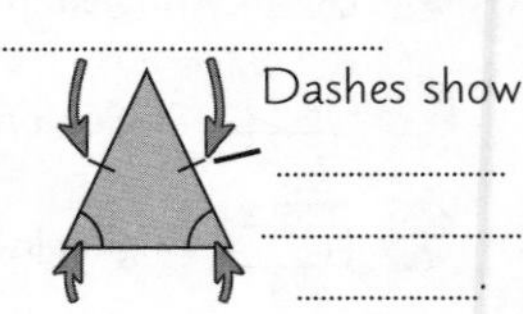

Angles Around Parallel Lines

When a line :

- Two bunches of
- There are
-

Arrows show

Alternate Angles

Found in a Z-shape:

Alternate angles .

Corresponding Angles

Found in an F-shape:

Corresponding angles .

Allied Angles

Found in a C- or U-shape:

Allied angles .

...... + ... = ... °

Polygons

First Go: ..... / ..... / .....

Regular Polygons

Name	Pentagon			Octagon	Nonagon	
No. of sides		6	7	8		10

Number of lines of symmetry = Number of .... = Order of .... symmetry

Interior and Exterior Angles

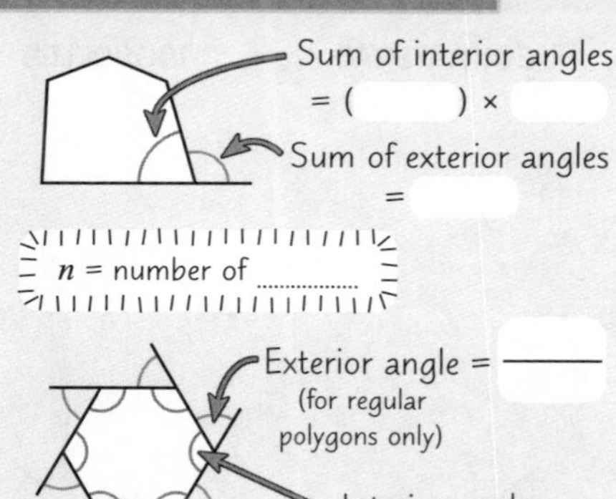

Four Types of Triangles

EQUILATERAL

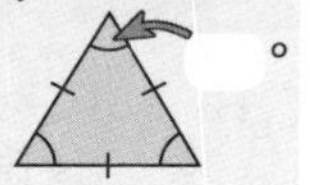

- lines of symmetry
- Rotational sym. order

ISOSCELES

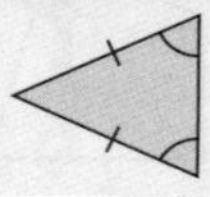

- of symmetry
- rotational symmetry

RIGHT-ANGLED

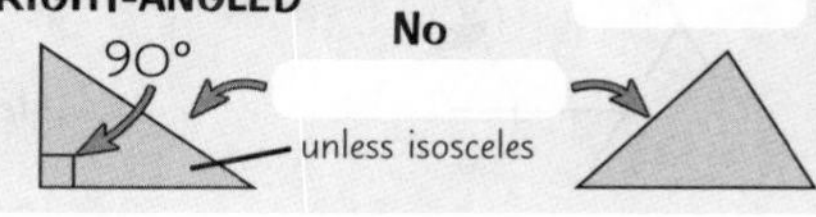

No rotational symmetry = order ....

Six Types of Quadrilaterals

SQUARE

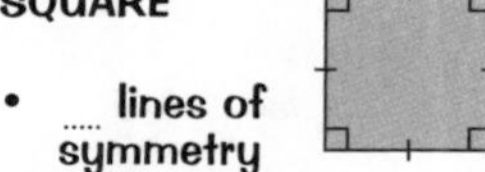

- ..... lines of symmetry
- Rotational sym. order .....
- Diagonals ...., cross at right angles

RECTANGLE

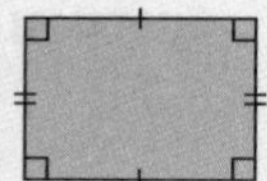

- lines of symmetry
- Rotational sym. order
- Diagonals

RHOMBUS

Add up to ....°

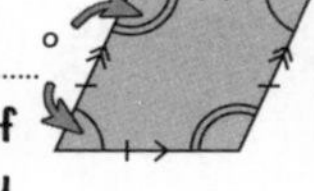

- ..... lines of symmetry
- Rotational sym. order .....
- .... cross at right angles

PARALLELOGRAM

Add up to °

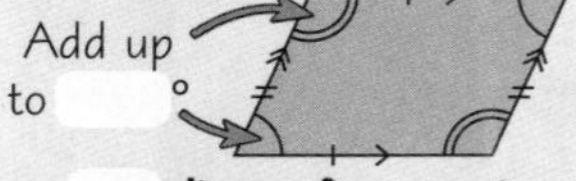

- lines of symmetry
- Rotational sym. order

TRAPEZIUM

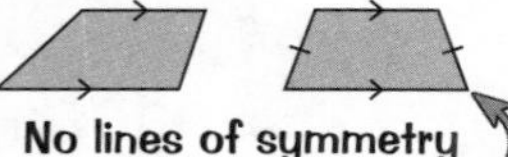

- No lines of symmetry (unless ....)
- .... rotational symmetry

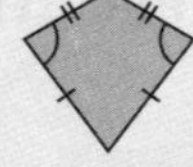

- line of symmetry
- rotational symmetry
- cross at right angles

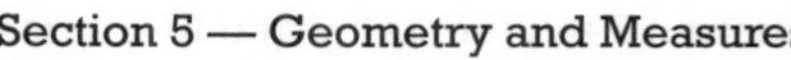

Second Go: ..... /..... /.....	Polygons

Polygons

Regular Polygons

Name	**Pentagon**			**Octagon**		
No. of sides			**7**			

Number of lines of .... **= Number of** .... **= Order of** ....

Interior and Exterior Angles

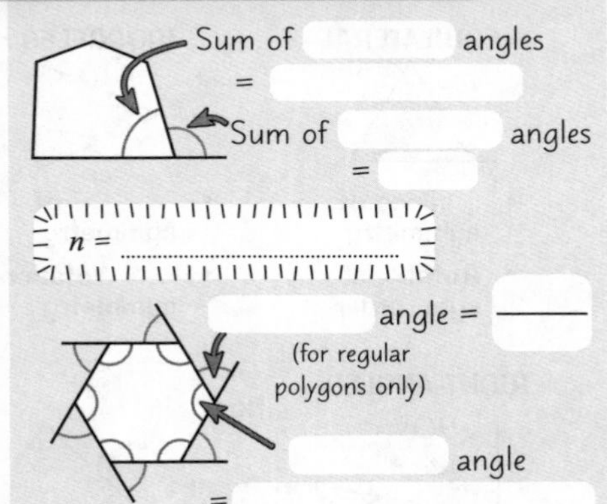

Four Types of Triangles

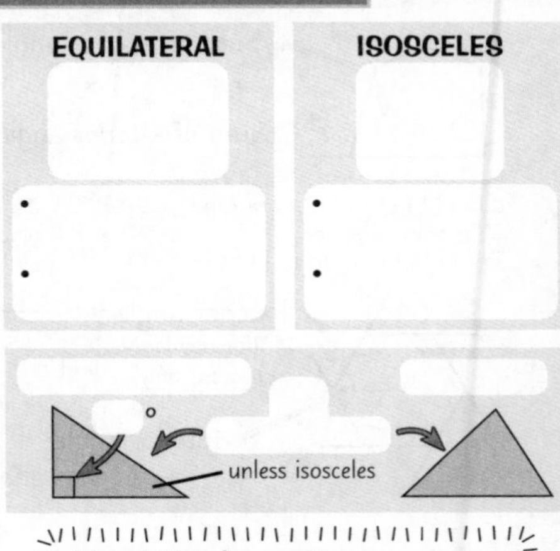

No rotational symmetry = ...

Six Types of Quadrilaterals

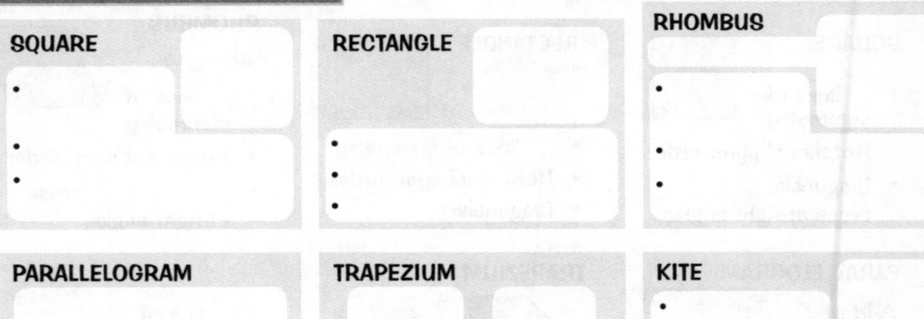

Circle Geometry

First Go: ..... / ..... / .....

Three Rules with Tangents

1. A tangent and a radius meet at ___°.
2. Tangents from the same point are the ___.

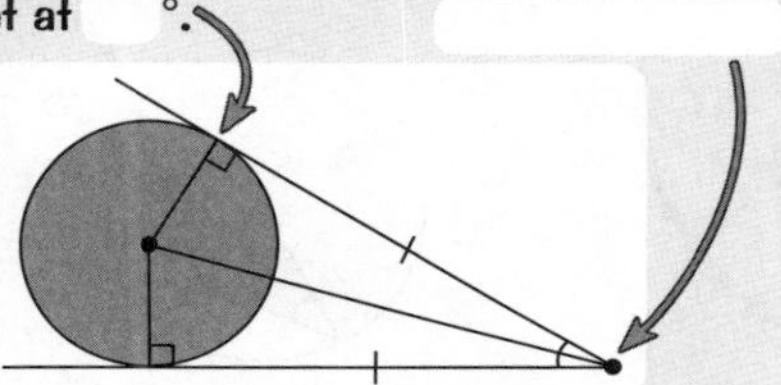

3. ALTERNATE SEGMENT THEOREM
The angle between a tangent and a ___ is ___ to the angle in the opposite segment.

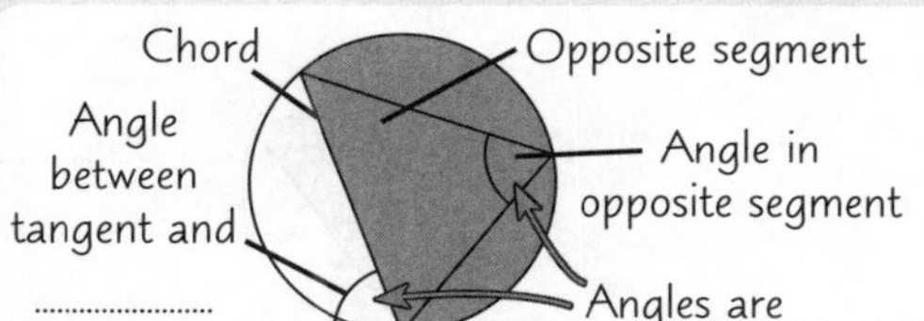

Two Rules with Polygons

1. Two radii form an ___ triangle.

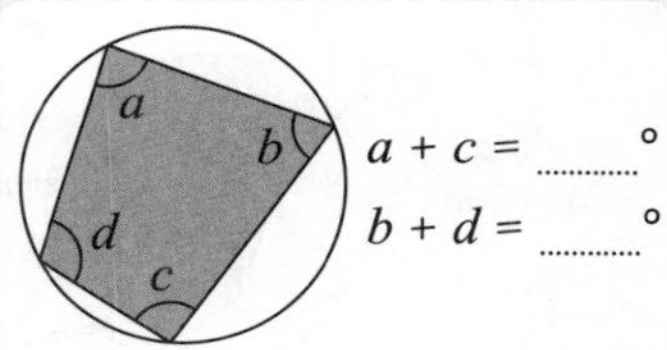

2. Opposite angles in a cyclic quadrilateral add up to ___°.

$a + c =$ .... °

$b + d =$ .... °

Four More Rules

1. The perpendicular ___ of a chord passes through the ___.

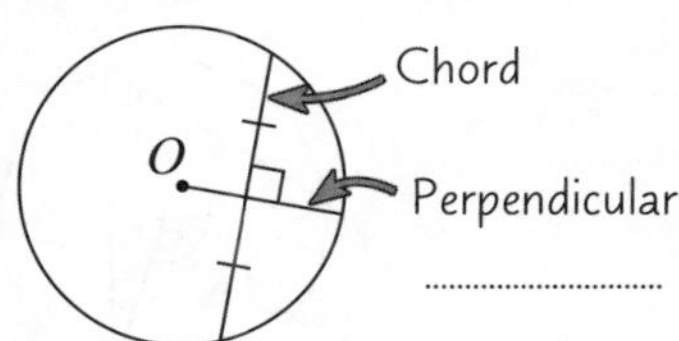

2. Angle made at the centre is ___ the angle made at the circumference.

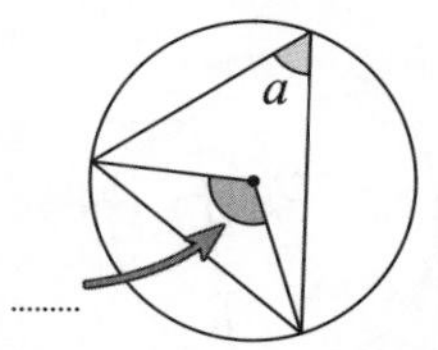

3. Angle in a ___ is 90°.

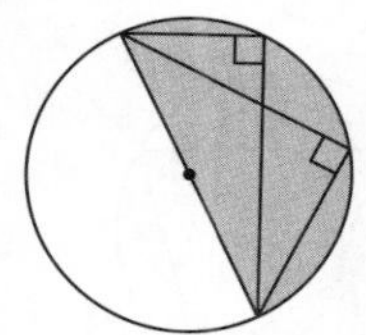

Two angles in opposite segments add up to ....°.

4. Angles in the same segment are ___.

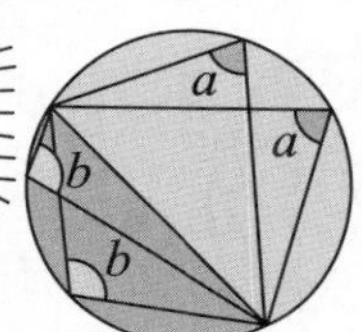

$a + b =$ .... °

Second Go:
...... /...... /......

Circle Geometry

Three Rules with Tangents

1

2

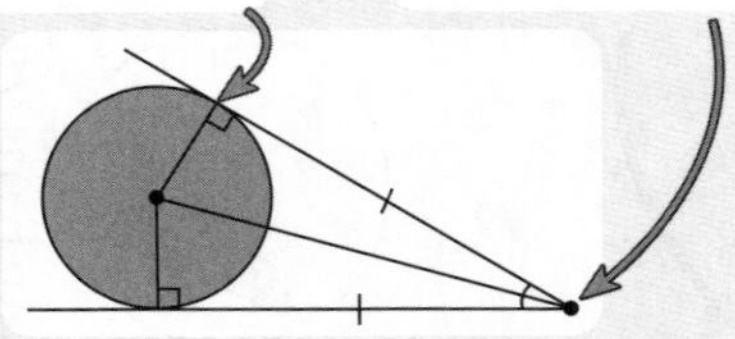

3 **ALTERNATE SEGMENT THEOREM**

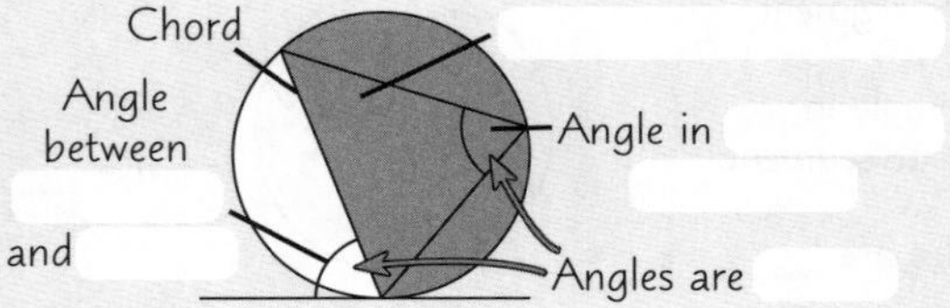

Chord

Angle between

and

Angle in

Angles are

Two Rules with Polygons

1 **Two radii**

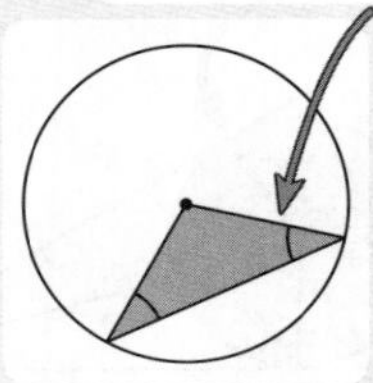

The radii are
....
....
....

2

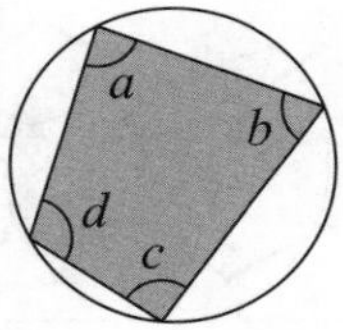

a b c d
....
....

Four More Rules

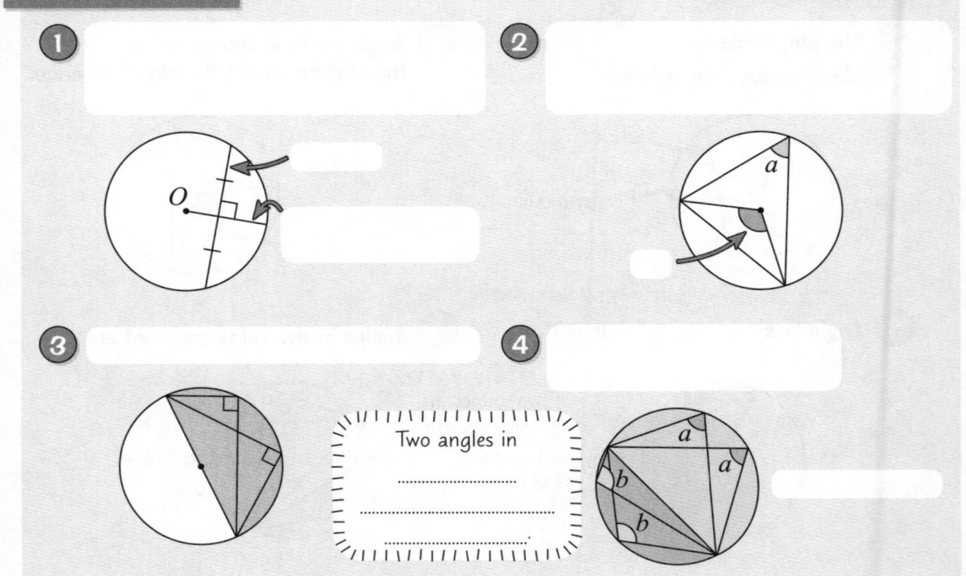

Mixed Practice Quizzes

Have a go at these quizzes — they cover what you've practised on p.85-90. They've shaped up quite nicely (if I do say so myself), so get ready for a party.

Quiz 1

Date: ...... / ...... / ......

1) Angles a and b are in the same segment. Angle a = 70°. What is the size of angle b?
2) One angle in a rhombus is 100°. What are the sizes of the other angles?
3) What do the exterior angles of a polygon add up to?
4) Angles a and b are corresponding. Angle a = 80°. What is the size of angle b?
5) How many lines of symmetry does an isosceles triangle have?
6) True or false? Angles around a point add up to 360°.
7) OA and OB are both radii of the same circle. What type of shape is OAB?
8) How many sides does a heptagon have?
9) What is the order of rotational symmetry of a parallelogram?
10) True or false? Opposite angles in a cyclic quadrilateral add up to 360°.

Total:

Quiz 2

Date: ...... / ...... / ......

1) True or false? Allied angles add up to 180°.
2) C is a point on the circumference of a circle with diameter AB. What is the size of angle ACB?
3) Which type of triangle has rotational symmetry of order 3?
4) A quadrilateral has three angles that add up to 300°. What is the size of the fourth angle?
5) What is the size of the angle made where a tangent and radius meet?
6) What do angles on a straight line add up to?
7) Which angle is equal to the angle between a tangent and a chord?
8) Angles a and c are opposite angles in a cyclic quadrilateral. What is a + c?
9) What is the formula for the sum of interior angles in a polygon?
10) AB and AC are both tangents to a circle. B and C are on the circumference. The length of AB is 6 cm. What is the length of AC?

Total:

Mixed Practice Quizzes

Quiz 3

Date: ...... / ...... / ......

1) How many identical angles does an isosceles triangle have?
2) True or false? A radius that crosses a chord at 90° splits the chord into two equal pieces.
3) Which type of quadrilateral has 1 line of symmetry and diagonals that cross at right angles?
4) What type of angles are found in a Z-shape?
5) True or false? A trapezium has rotational symmetry of order 2.
6) The exterior angle of a regular polygon is 45°. What is the size of each interior angle?
7) What is the alternate segment theorem?
8) a and b are allied angles. Angle a = 30°. What is the size of angle b?
9) What is the name of a nine-sided polygon?
10) How much bigger is the angle made at a circle's centre than the angle made at its circumference?

Total:

Quiz 4

Date: ...... / ...... / ......

1) Which type of quadrilateral has 4 lines of symmetry?
2) What is the size of the angle made in a semicircle?
3) What do the angles in a triangle add up to?
4) OA is a radius of a circle. AB is a tangent. What is the size of angle OAB?
5) Angles a and b are vertically opposite. Angle a = 50°. What is the size of angle b?
6) What do angles in opposite segments add up to?
7) True or false? Corresponding angles are equal.
8) What does the number of sides tell you about a regular polygon's rotational symmetry?
9) Points A and B are on the circumference of a circle with centre O. The length of OA is 3 cm. What is the length of OB?
10) What is the size of an exterior angle of a regular hexagon?

Total:

Congruent and Similar Shapes

First Go: ..... / ..... / .....

Four Conditions for Congruent Triangles

Shapes are congruent under translation, ... and reflection.

CONGRUENT — .... size and .... shape.

Condition	1 SSS	2 ASA	3	4 RHS
Description	three ______ the same	two ______ and corresponding side match up	two sides and angle ______ them match up	______ angle, hypotenuse and another side all match up
Diagrams				

Two Steps to Prove Congruence

1 .... down everything you know.

State which .... holds and why.

EXAMPLE

O is the centre of this circle. Prove that triangles *ABO* and *CDO* are congruent.

1. *AO*, *BO*, *CO* and *DO* are all radii, so they're ______. Angles *AOB* and ______ are vertically opposite, so they're equal.
2. ______ — two sides and the angle between them match up, so *ABO* and *CDO* are ______.

Three Conditions for Similar Triangles

Shapes are .... under enlargement.

SIMILAR — ______ shape, ______ size.

1 All ______ match up.

All sides are ______.

3 Two sides proportional and ______ between is the same.

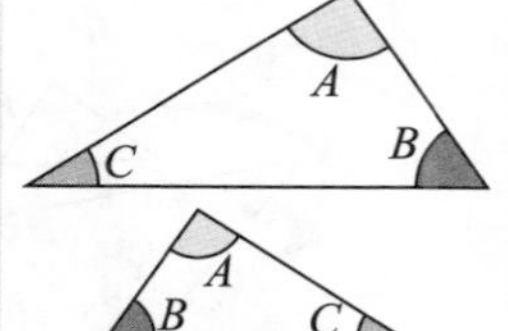

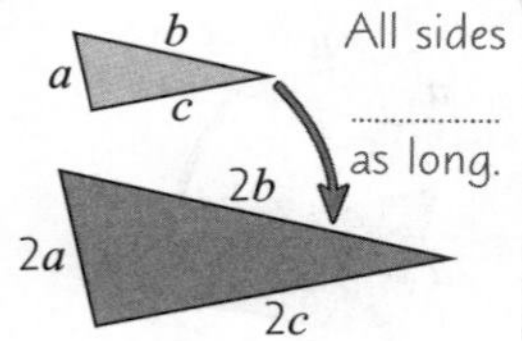

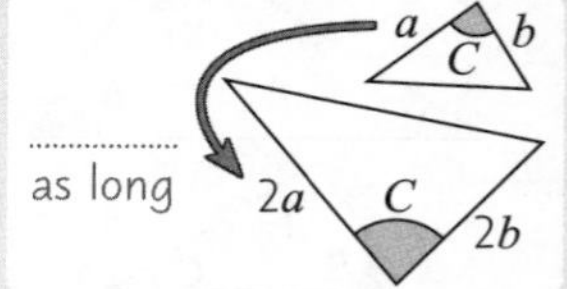

Second Go:
...../...../.....

Congruent and Similar Shapes

Four Conditions for Congruent Triangles

Shapes are ....
....
....

CONGRUENT — ....

Condition	1	2	3	4
Description				
Diagrams	b, c, a / a, c, b	A, B, c / B, A, c	c, B, a / c, B, a	b, c, A / c, A, b

Two Steps to Prove Congruence

1 **Write** ....

2 **State** ....

EXAMPLE

O is the centre of this circle. Prove that triangles *ABO* and *CDO* are congruent.

1 ...., so they're equal.
....
..............., so they're equal.

2 .... — ....
..............., so *ABO* and *CDO* are congruent.

Three Conditions for Similar Triangles

Shapes are ....
....

SIMILAR — ....

1

2

....
....
....

3

....
....

The Four Transformations

First Go: ...../...../.....

Translation

Amount a shape moves is given by $\begin{pmatrix} x \\ y \end{pmatrix}$.

x = ... movement

y = ... movement

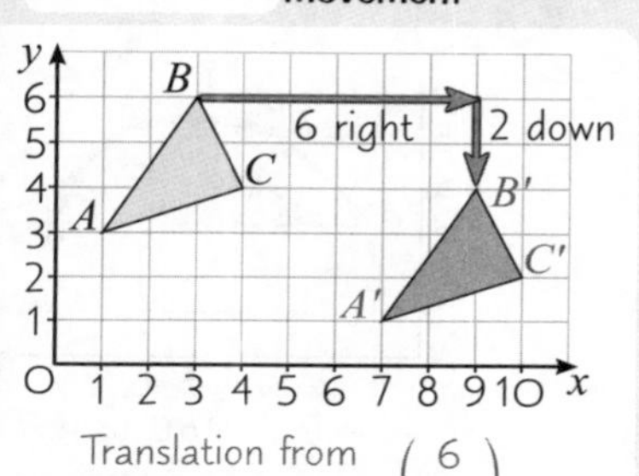

Translation from ABC to $A'B'C'$: $\begin{pmatrix} 6 \\ ...... \end{pmatrix}$

Rotation

To describe a rotation you need:

1. the angle
2. the ....
3. the .... of rotation

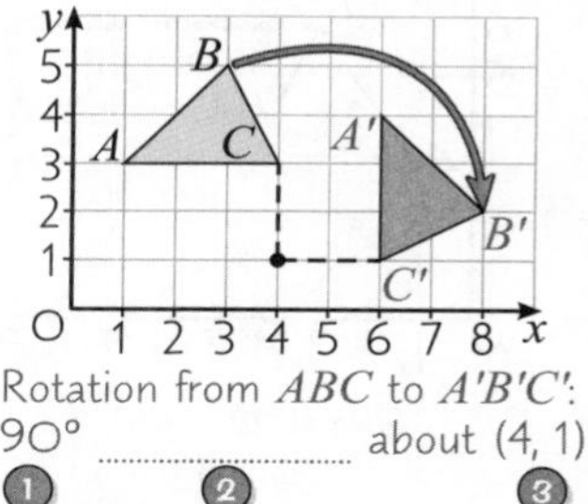

Rotation from ABC to $A'B'C'$:
90° (1) .... (2) about (4, 1) (3)

Reflection

Describe by giving the equation of the ....

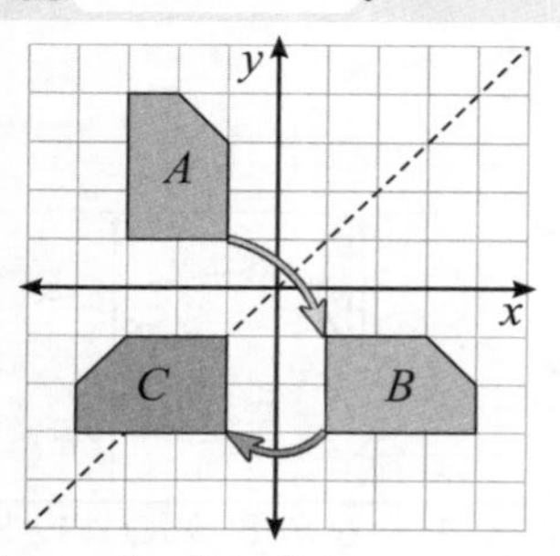

B is a reflection of A in ....

C is a reflection of ..... in the y-axis

Enlargement

To describe an enlargement you need:

1. the .... $= \dfrac{\text{new length}}{\text{old length}}$
2. the .... of enlargement

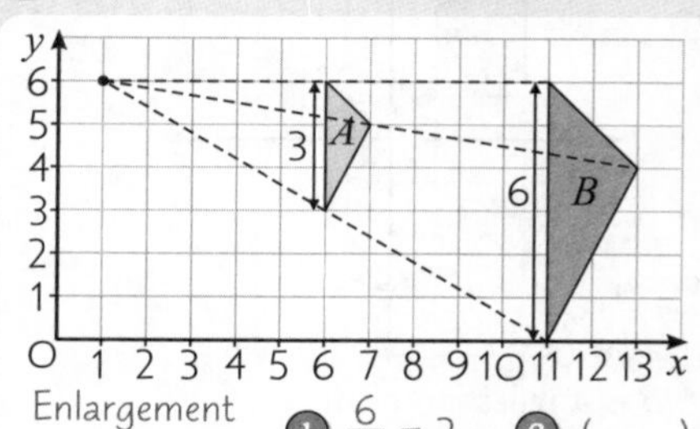

Enlargement from A to B: (1) $\dfrac{6}{3} = 2$ (2) (....., ...)

Four Facts about Scale Factors

1. If .... than 1, shape gets bigger.
2. If .... than 1, shape gets smaller.
3. If ...., shape goes to other side of centre of enlargement. Scale factor of ...... = rotation of 180°.
4. They give relative distance of new and old points from ....

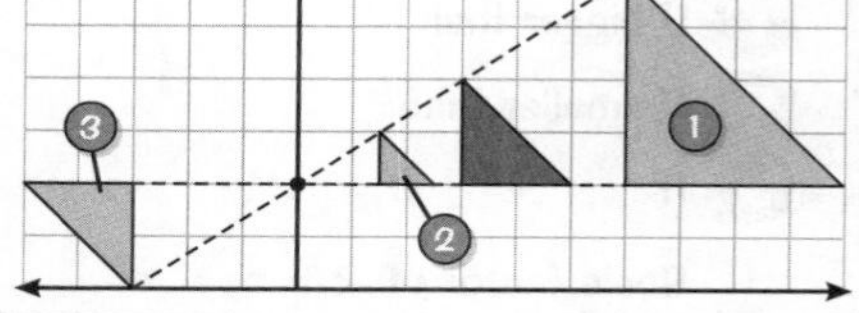

Second Go:
...../...../.....

The Four Transformations

Translation

Amount a shape moves is given by .

x =

y =

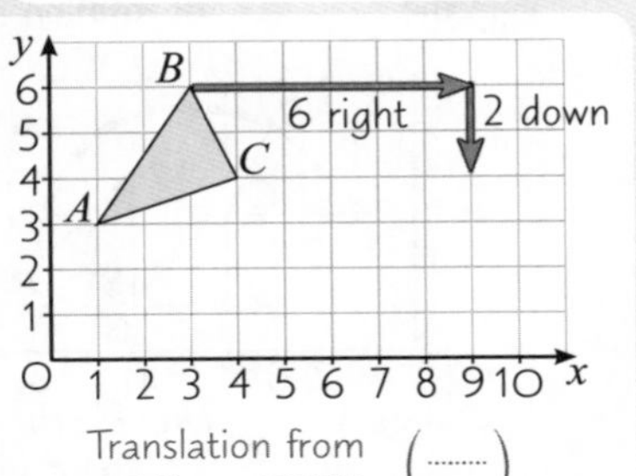

Rotation

To describe a rotation you need:

1 2

3

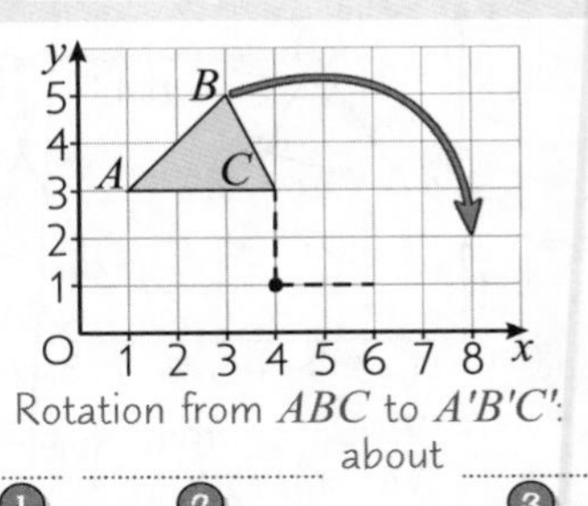

Reflection

Describe by giving the .

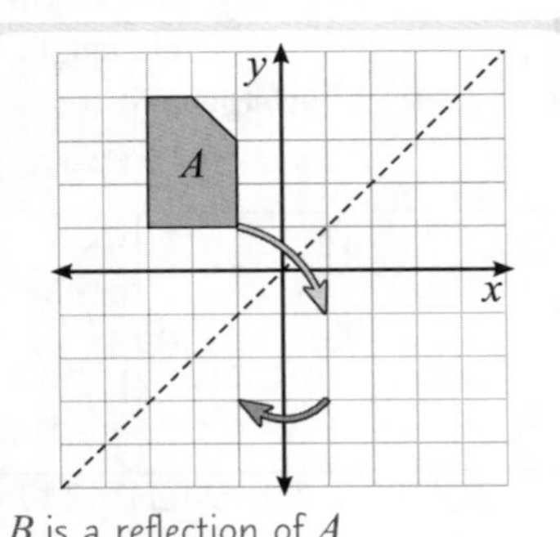

Enlargement

To describe an enlargement you need:

1

2

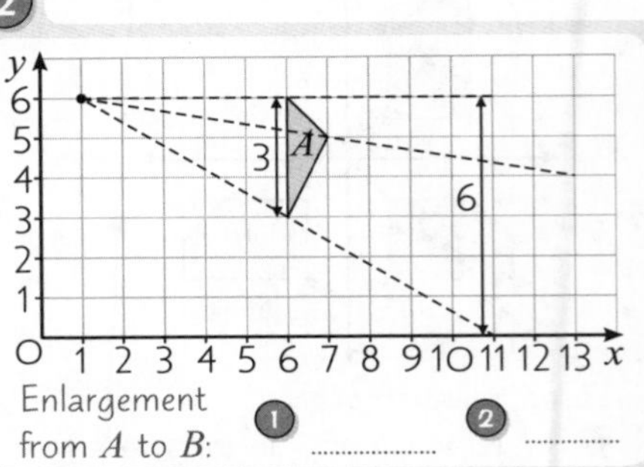

Four Facts about Scale Factors

1. If bigger than
2. If smaller than
3. If

   Scale factor of –1 =
4. They give

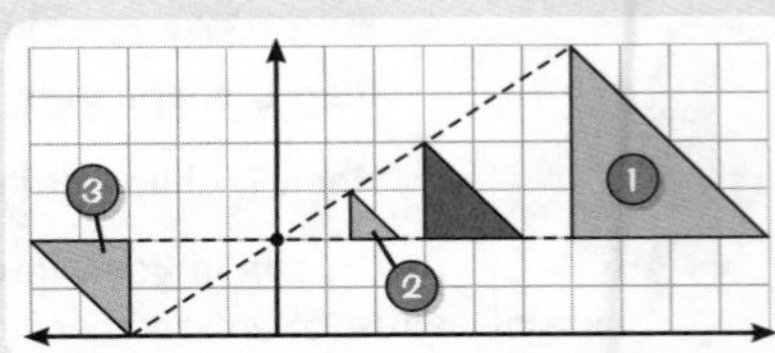

Perimeter and Area

First Go: ...../...../.....

Triangles and Quadrilaterals

Area of rectangle = ________ × width

Squares have equal length and width so area = ...

Area of triangle = $\frac{1}{2}$ × ______ × vertical height

Area of parallelogram = ______ × vertical height

Area of trapezium = $\frac{1}{2}$(______) × vertical height

Split composite shapes into triangles and quadrilaterals. Work out each area and ____________.

5 cm² | 10 cm² | Total area: 15 cm²

Only include ____________ when adding up perimeters.

Circles

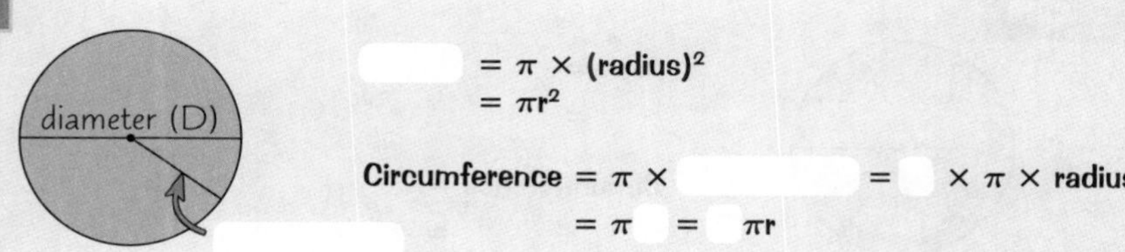

______ = π × (radius)² = πr²

Circumference = π × ________ = ___ × π × radius = π___ = ___πr

Arcs and Sectors

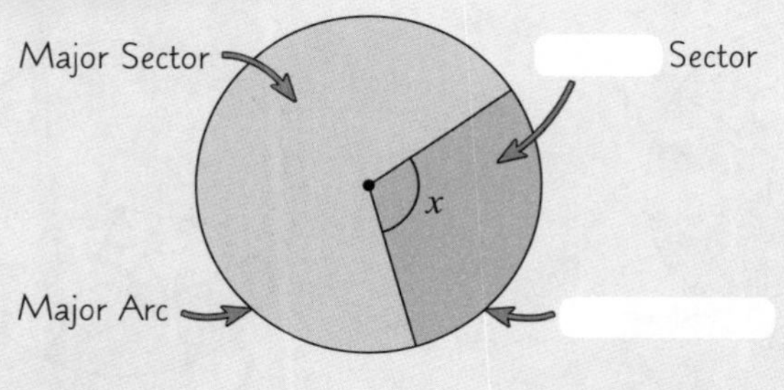

Area of ______ = $\frac{x}{360}$ × ______ of full circle

Length of ______ = $\frac{x}{360}$ × circumference of full circle

Segments

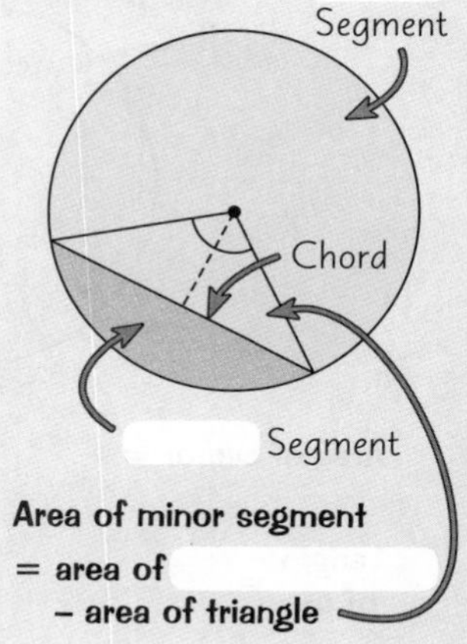

Area of minor segment = area of ____________ – area of triangle

Second Go:
...../...../.....

Perimeter and Area

Triangles and Quadrilaterals

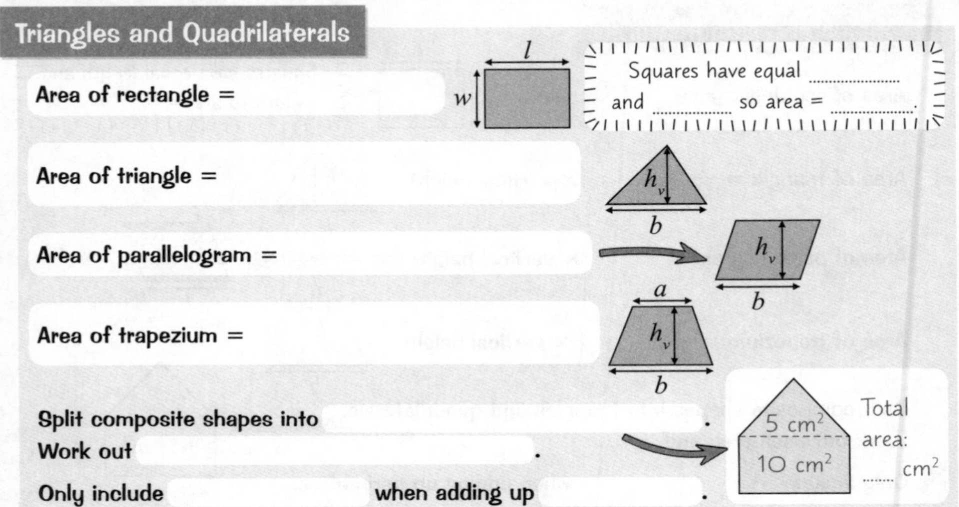

Area of rectangle =

Area of triangle =

Area of parallelogram =

Area of trapezium =

Split composite shapes into .
Work out .
Only include when adding up .

Circles

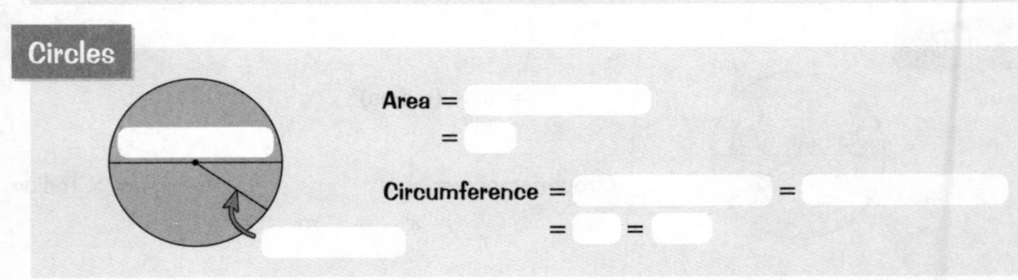

Area =
=

Circumference = =
= =

Arcs and Sectors

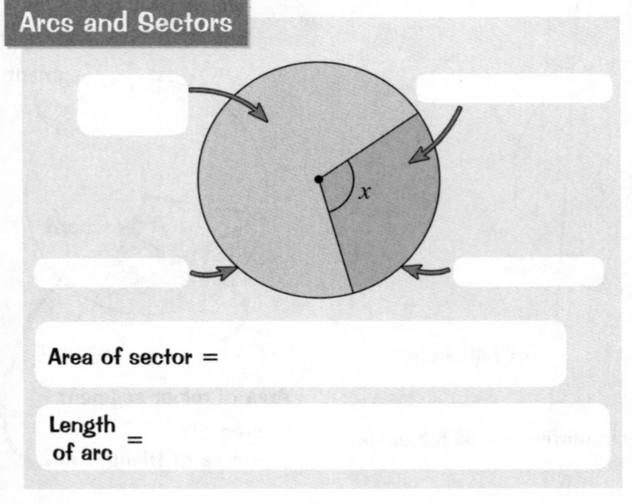

Area of sector =

Length of arc =

Segments

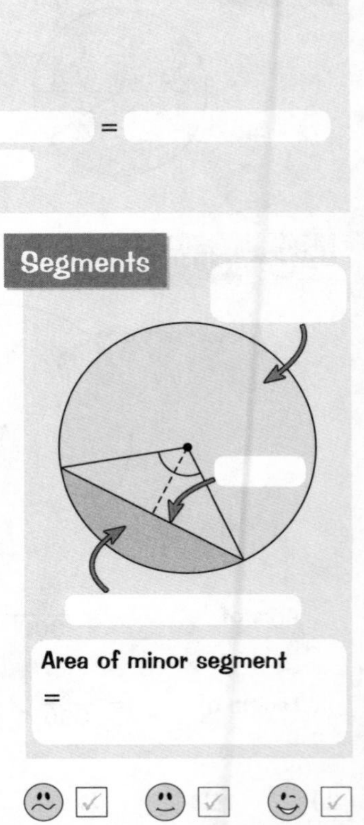

Area of minor segment
=

Mixed Practice Quizzes

Congratulations — you're the lucky winner of four quizzes on the topics covered on p.93-98 (terms and conditions may apply...). Aren't you lucky.

Quiz 1

Date: ...... / ...... / ......

1) A chord splits a circle into two parts. What is the larger part called?
2) True or false? You can tell that two triangles are congruent if two sides and any angle match up.
3) What effect does a negative scale factor have on an enlargement?
4) What do you use the formula πr^2 to find?
5) What are the coordinates of the point (2, 3) when reflected in the y-axis?
6) What does the congruence condition RHS mean?
7) Two triangles are similar if all their sides are what?
8) What is the formula for the scale factor of an enlargement?
9) What is the area of a triangle with base 2 cm and vertical height 3 cm?
10) What three pieces of information do you need to describe a rotation?

Total:

Quiz 2

Date: ...... / ...... / ......

1) What happens to a shape when it is enlarged by a scale factor of −2?
2) For the same angle in the same circle, which has the larger area: the minor sector or minor segment?
3) Under which type of transformation are shapes similar but not congruent?
4) How do you work out the area of a composite shape?
5) What is the formula for the length of an arc?
6) What are the four different conditions for congruent triangles?
7) In a vector describing a translation, does the top or bottom number show the vertical movement?
8) Is knowing that all of the angles match up in two triangles enough to prove that they are: a) similar? b) congruent?
9) A shape is rotated 90° with a centre of (2, 3). What else do you need to fully describe the rotation?
10) What is the formula for the area of a parallelogram?

Total:

Mixed Practice Quizzes

Quiz 3

Date: ...... / ...... / ......

1) What two pieces of information do you need to describe an enlargement?
2) Two triangles are the same size and shape. Are they congruent or similar?
3) In a translation, a shape moves 3 units to the right and 2 units down. What vector describes the translation?
4) What is the formula for the area of a sector?
5) How many different pieces of information do you need to show that two triangles are congruent?
6) What scale factor has the same effect as a rotation of 180°?
7) Which parts of a composite shape do you use to find the perimeter?
8) True or false? Two triangles are similar if two sides are proportional and the angle between them is the same.
9) What is the formula for the area of a triangle?
10) A'B'C' is an enlargement of ABC. If AB is 4 cm and A'B' is 12 cm, what is the scale factor?

Total:

Quiz 4

Date: ...... / ...... / ......

1) What information do you need to describe a reflection?
2) True or false? Shapes are congruent under rotation.
3) What is the formula for the area of a trapezium?
4) How do you find the area of a minor segment?
5) Triangles ABC and DEF are similar. AB is 1 cm, BC is 2 cm, AC is 2.5 cm, DE is 3 cm, and EF is 6 cm. What is the length of DF?
6) The scale factor for an enlargement is 0.25. What happens to the shape?
7) What is the formula for the circumference of a circle?
8) If you know that two angles match in each of two triangles, what else do you need to show they are congruent?
9) (4, 1) is on shape A. The enlargement of A to B has scale factor 3 and centre of enlargement (0, 1). What are the coordinates of the point on B?
10) Two radii split a circle into two pieces. What is the bigger piece called?

Total:

3D Shapes and Surface Area

First Go: ..... /..... /.....

Parts of 3D Shapes

If you're asked to find the number of vertices/edges/faces, just them up — don't forget hidden ones.

E.g. this cube has 8 vertices, ... edges and ..... faces.

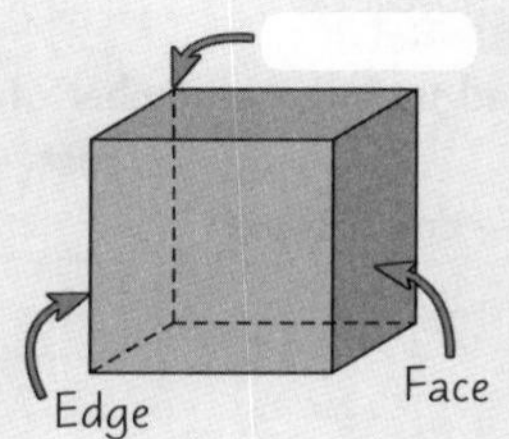

Three Projections

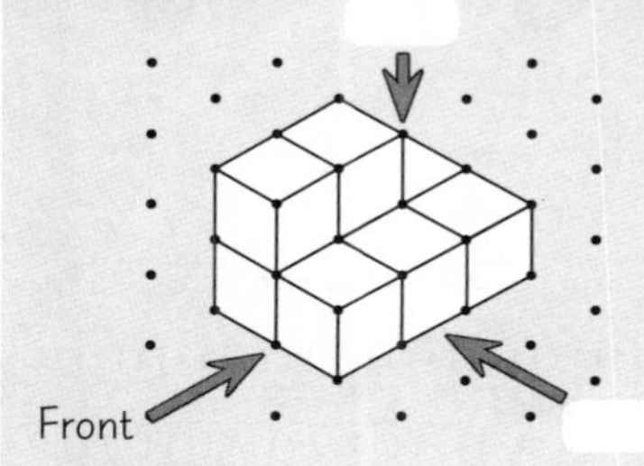

1 Front elevation

2 elevation

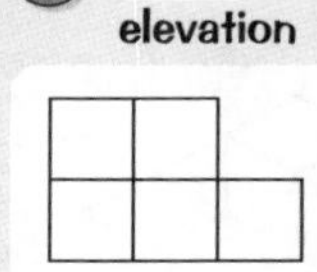

3

Dotty paper is called .... paper.

Surface Area

SURFACE AREA — total area of all .

Surface area of solid = of net

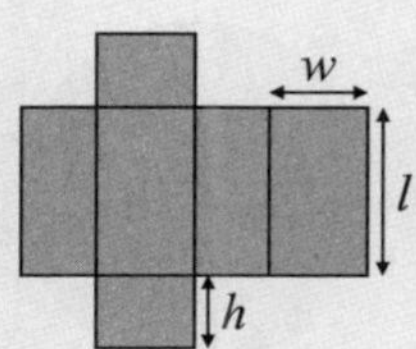

Surface area of sphere =

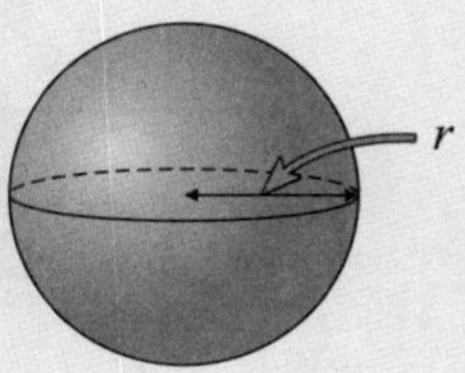

Surface area of cone = πrl +

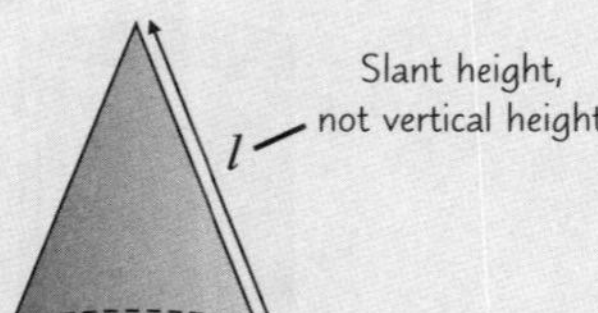

Surface area of cylinder = + $2\pi r^2$

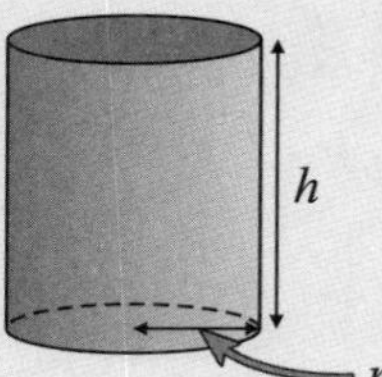

Second Go:
..... /..... /.....

3D Shapes and Surface Area

Parts of 3D Shapes

If you're asked to find the number of ______________, just ______ ______ — don't forget ______________.

E.g. this cube has ....
....

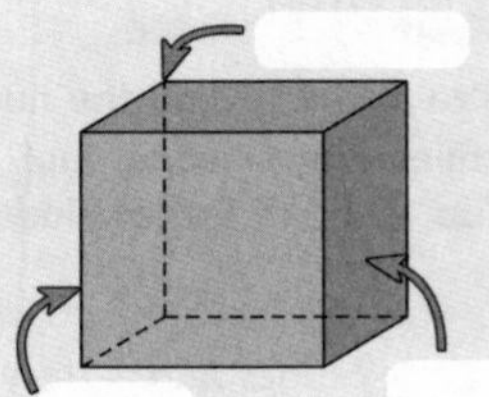

Three Projections

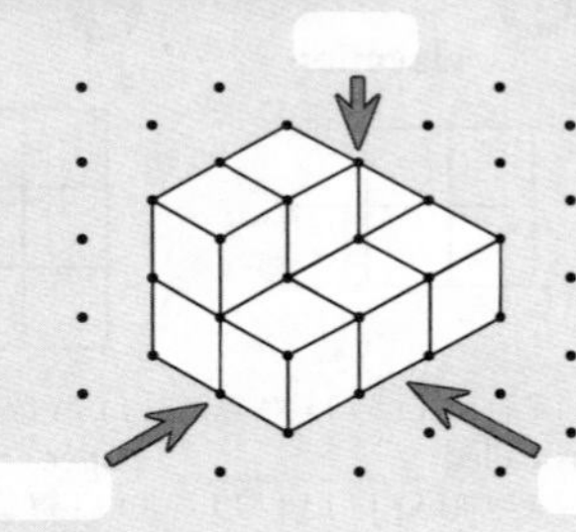

1

2

3

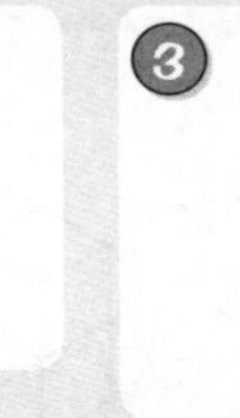

Surface Area

SURFACE AREA — ______________________.

Surface area of solid =

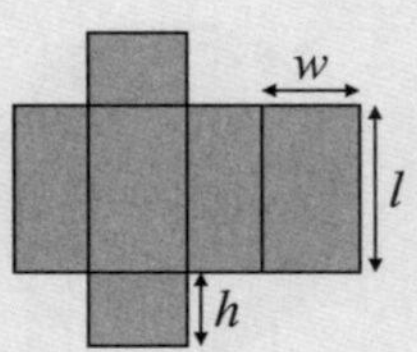

Surface area of sphere =

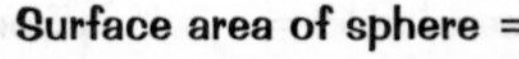

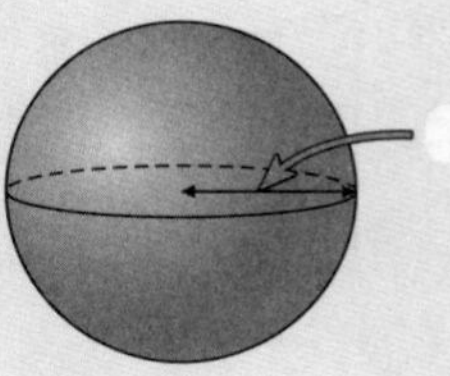

Surface area of cone =

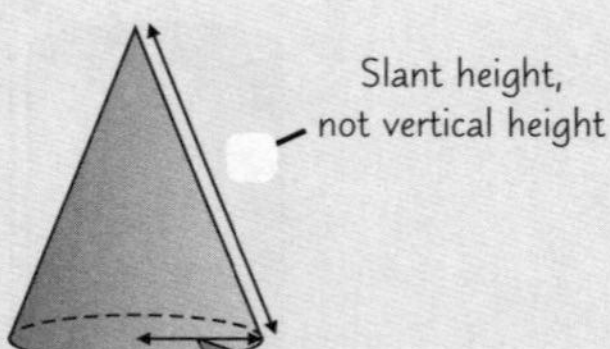

Surface area of cylinder =

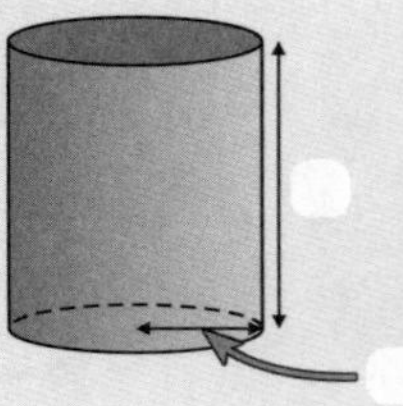

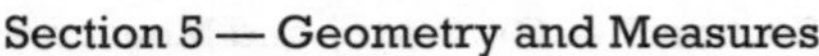

Volume and Enlargement

First Go:
...../...../.....

Six Volume Formulas

1. Volume of prism
= A ×

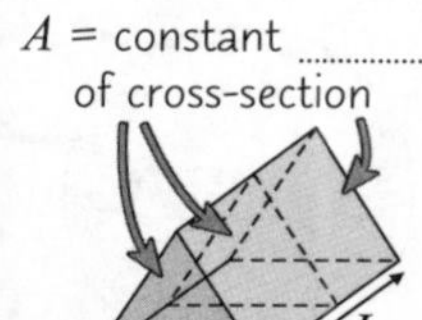

2. Volume of cylinder
=

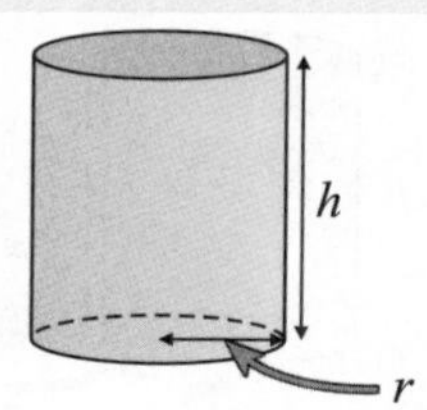

3. Volume of sphere
$= \frac{4}{3}\pi$

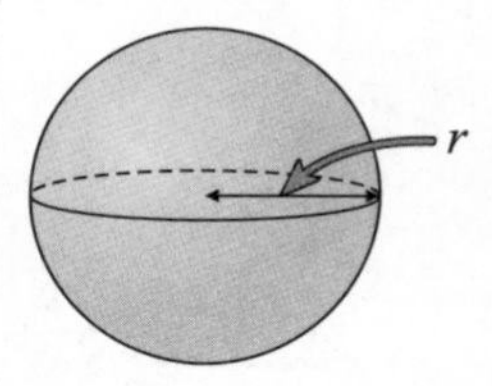

4. Volume of pyramid
$= \frac{1}{3}$ × base area ×

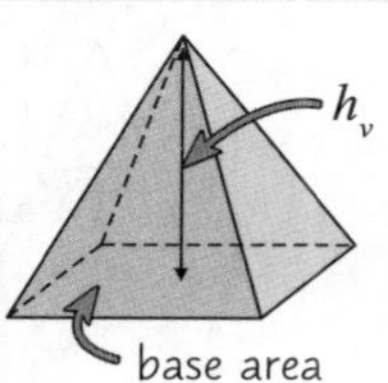

5. Volume of cone
$= \quad \pi r^2 h_v$

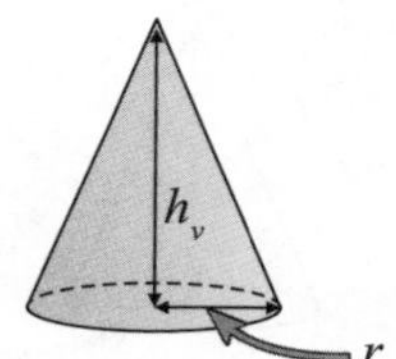

6. Volume of frustum
$= \frac{1}{3}\pi R^2 H - \frac{1}{3}\pi$

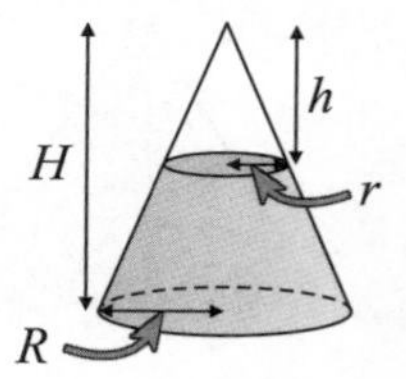

Rates of Flow

RATE OF FLOW — how fast .... is changing.

EXAMPLE

A cylinder with radius 10 cm and height 8 cm is filled with water at 1 litre per minute. How long does this take to the nearest second?

Find total volume:
$V = \pi \times \quad \times \quad = 2513.2...\ cm^3$

Convert units: 1 L = 1000 cm^3
1000 cm^3/min = 16.6...

Solve for time:
2513.2... ÷ = 151 s (to nearest s)

Enlargement of Areas and Volumes

If a shape changes by a scale factor of n:

Sides are times bigger (1:).

.... = $\frac{\text{new length}}{\text{old length}}$

Areas are times bigger (1:).

...... = $\frac{\text{new area}}{\text{old area}}$

Volumes are times bigger (1:).

...... = $\frac{\text{new volume}}{\text{old volume}}$

Second Go: ..... /..... /.....

Volume and Enlargement

Six Volume Formulas

1 Volume of prism

=

=

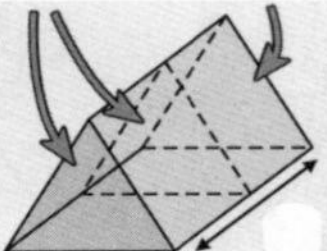

2 Volume of cylinder

=

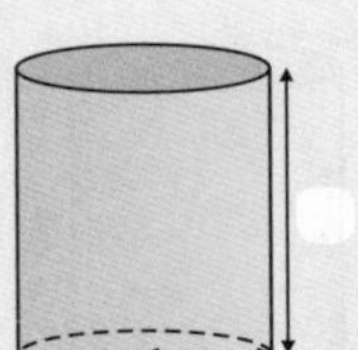

3 Volume of sphere

=

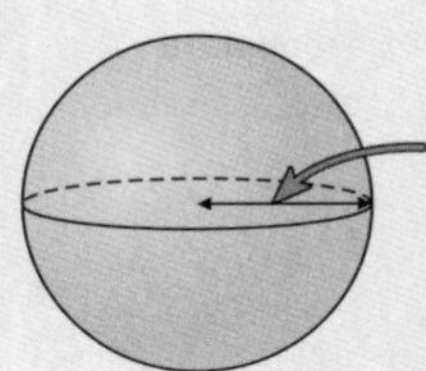

4 Volume of pyramid

=

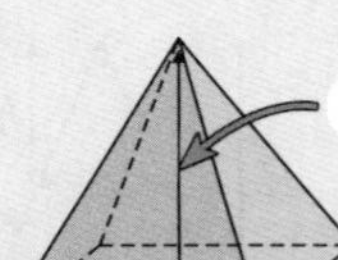

5 Volume of cone

=

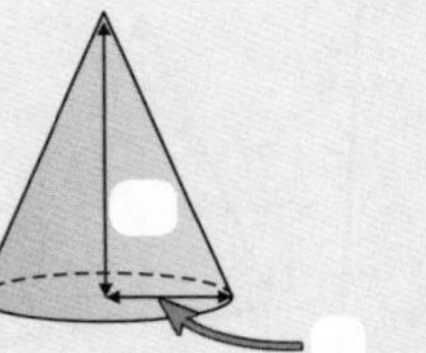

6 Volume of frustum

=

Rates of Flow

RATE OF FLOW —

EXAMPLE

A cylinder with radius 10 cm and height 8 cm is filled with water at 1 litre per minute. How long does this take to the nearest second?

Find total volume:

Convert units: 1 L = 1000 cm^3

Solve for time:

Enlargement of Areas and Volumes

If a shape changes by a
:

Sides are .

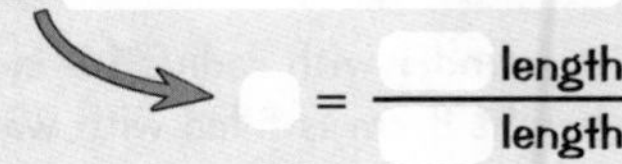

Areas are .

Volumes are .

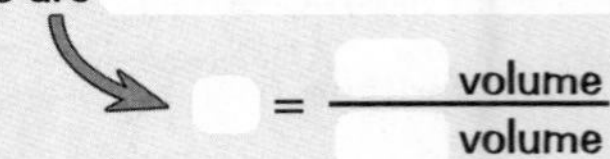

Triangle Construction

First Go:
..... / ..... / .....

Constructing Triangles

There's only ____ triangle you can draw if you're given:

SSS

ASA

There are .... triangles you could draw if you know two sides and an angle that isn't between them.

Four Steps for Three Known Sides

1. Roughly ____ and label the triangle.
2. Accurately draw and label the ____.
3. Set ____ to each side length, then draw an ____ at each end.
4. ____ the ends with the arc intersection. ____ points and sides.

EXAMPLE

Construct triangle ABC where AB = 3 cm, BC = 2 cm, AC = 2.5 cm.

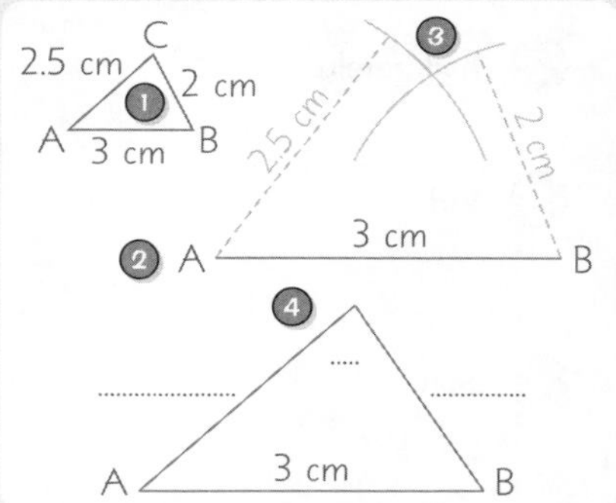

Five Steps for Known Sides and Angles

1. Roughly ____ and label the triangle.
2. Accurately draw and ____ the base line.
3. Use a ____ to measure the angles and mark out with ____.
4. **ASA** Draw lines from ends through .... Label the intersection.
   SAS Measure towards .... Label the point.
   RHS Draw line from one end through dot. Draw arc from .... Label the intersection.
5. ____ the points. Label known sides and angles.

EXAMPLE

Construct triangle XYZ where XY = 2 cm, angle YXZ = 70°, angle XYZ = 40°.

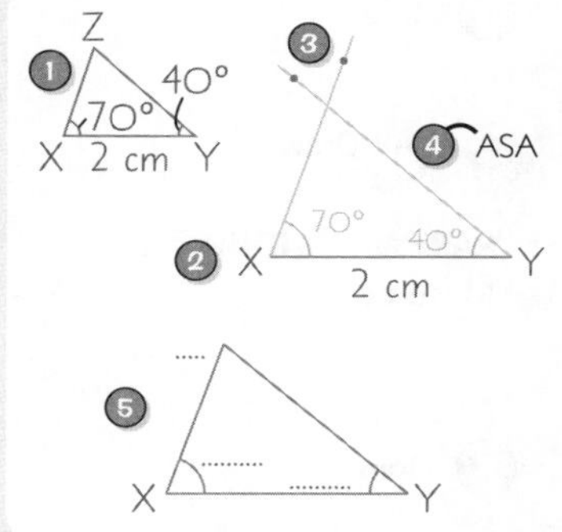

Second Go:
...../...../.....

Triangle Construction

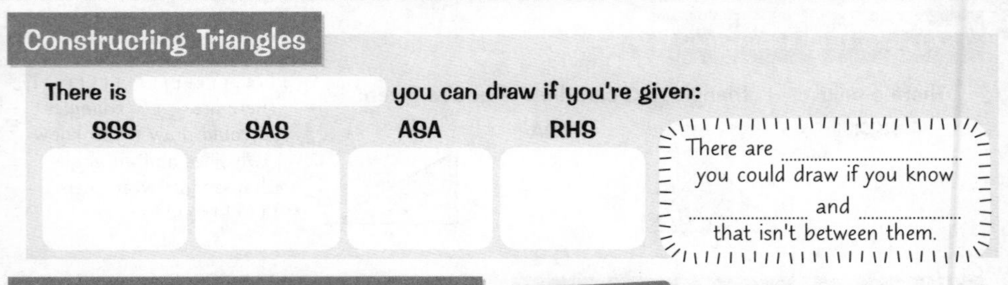

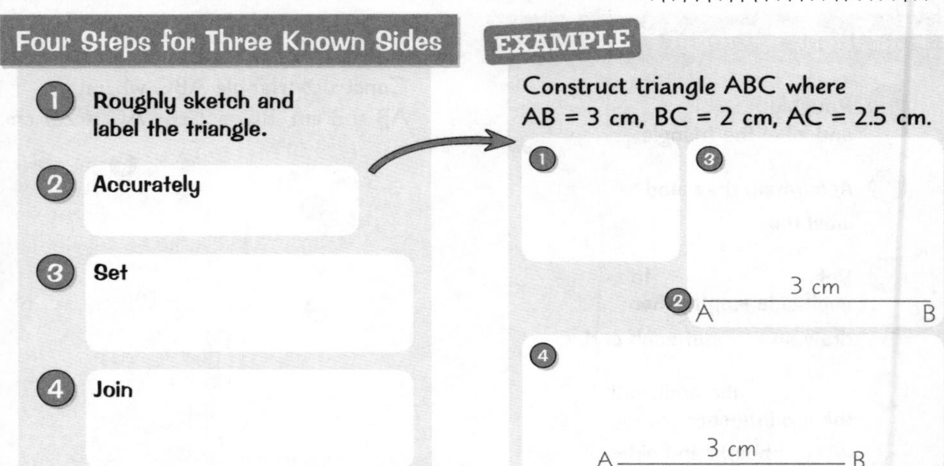

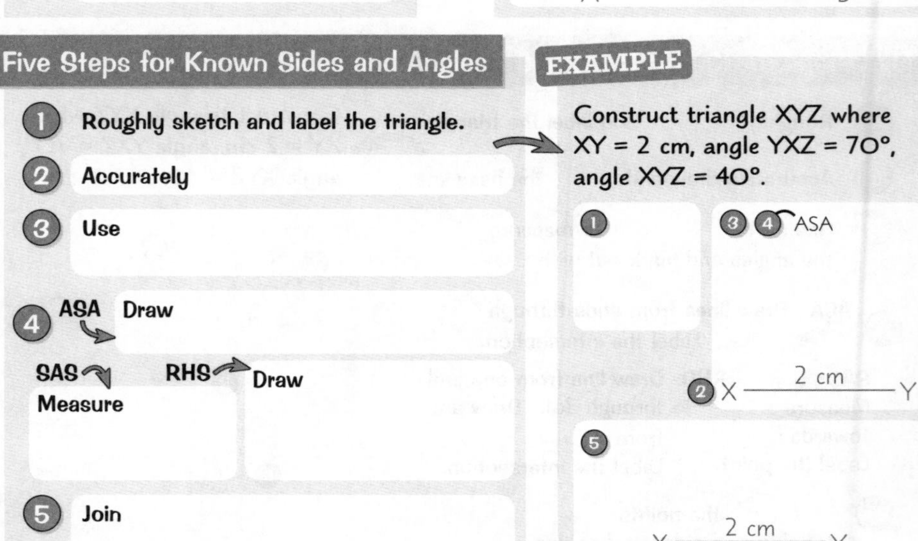

Loci and Construction

First Go: ...../...../.....

Four Different Types of Loci

LOCI — lines or regions showing that fit a given rule.

1. **Locus of points from a given point:**

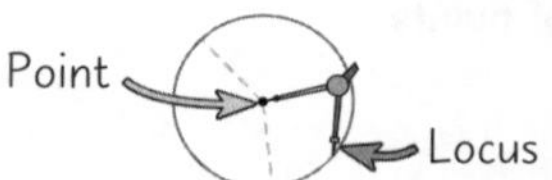

2. **Locus of points from a given line:**

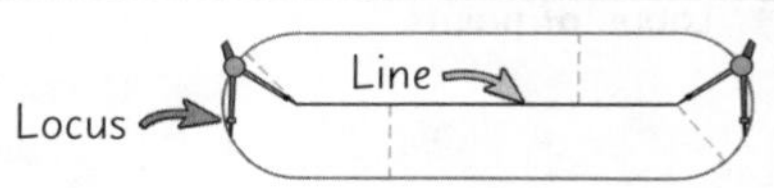

3. **Locus of points from two given lines:**

This locus .... the angle between the two lines.

4. **Locus of points from two given points:**

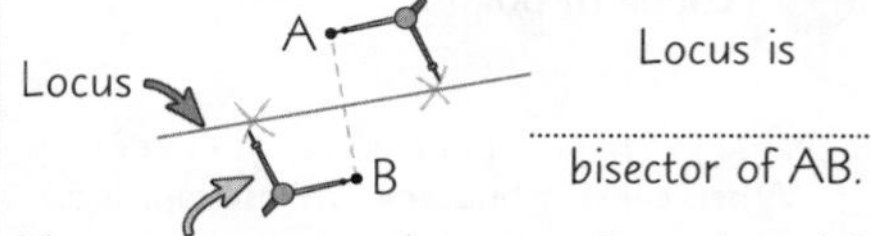

When constructing any of these four loci, keep your compass settings ....

Constructing 60° Angles

Keep compass settings for 60° angles.

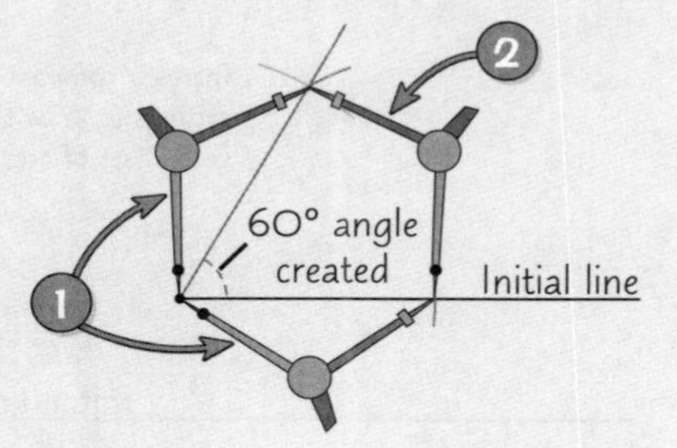

Constructing 90° Angles

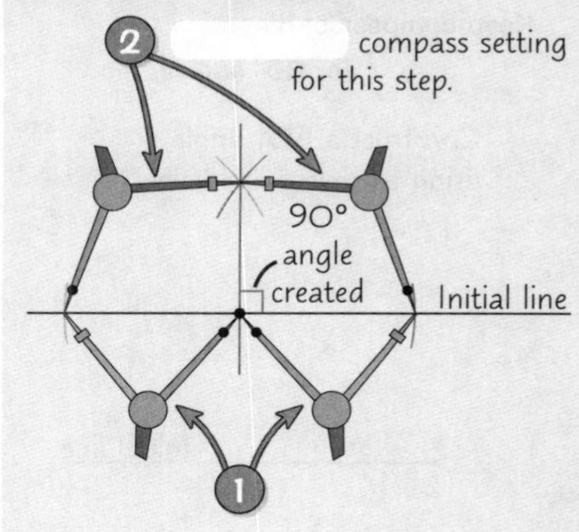

Second Go:
...../...../.....

Loci and Construction

Four Different Types of Loci

LOCI — .

1. **Locus of points**

2. **Locus of points**

3. **Locus of points**

This locus .... the two lines.

4. **Locus of points**

When constructing any of these four loci, ...

Constructing 60° Angles

Keep compass settings **for 60° angles.**

Construct a 60° angle using the initial line below.

Constructing 90° Angles

Construct a 90° angle using the initial line below.

Construction and Bearings

First Go: ..... /..... /.....

Drawing the Perpendicular From a Point to a Line

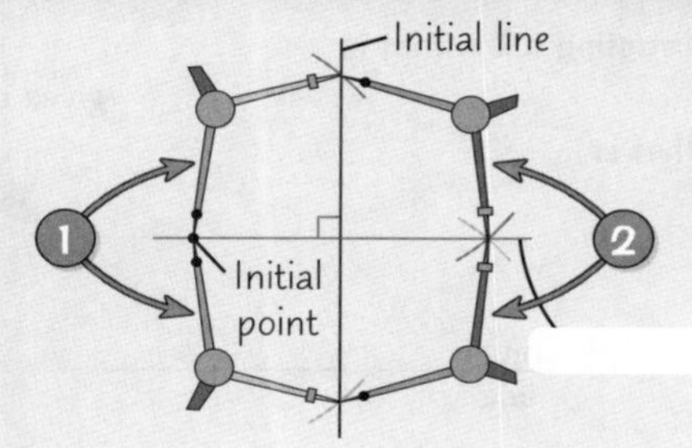

Constructing a line that is [blank] to the one you've just drawn gives a line that is parallel to the [blank].

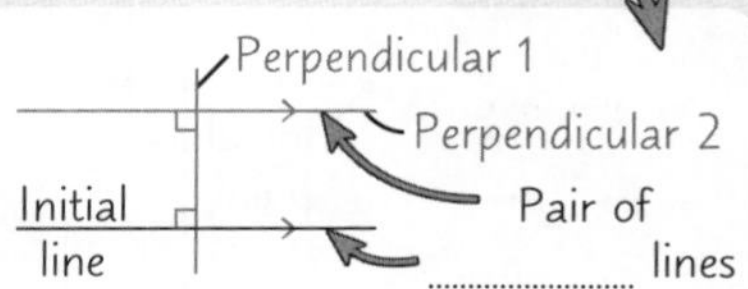

Three Steps to Find Bearings

1. Put your pencil at the point you're going [blank].
2. Draw a [blank] at that point.
3. Measure the angle [blank] from the [blank] to the line joining the two points.

Bearings must be given as ..... figures — e.g. ....° rather than 90°.

EXAMPLE

Find the bearing of X from Y.

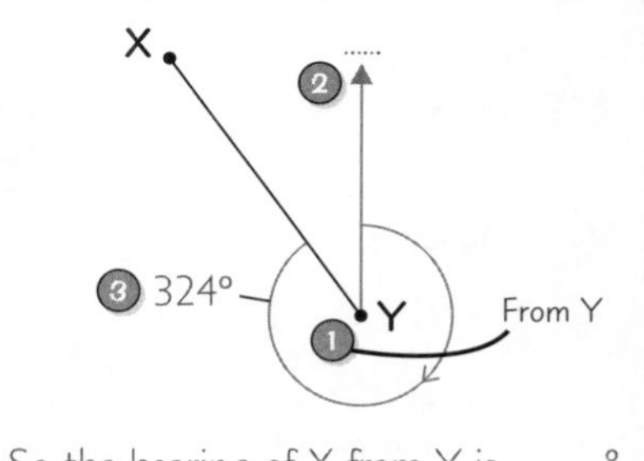

So the bearing of X from Y is ....°.

Scale Drawings

Scale drawings and maps show the positions of objects and the [blank] between them.

Real-life distance = [blank] × scale factor

Be wary of .... They're not usually the same for real life and the map.

Scale: 5 cm = 1 km. So .... = 0.2 km.

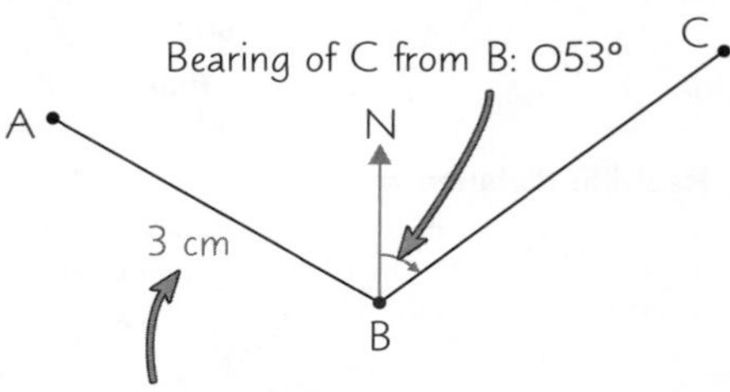

A to B in real life: 3 × .... = ... km

Second Go: ...../...../.....

Construction and Bearings

Drawing the Perpendicular From a Point to a Line

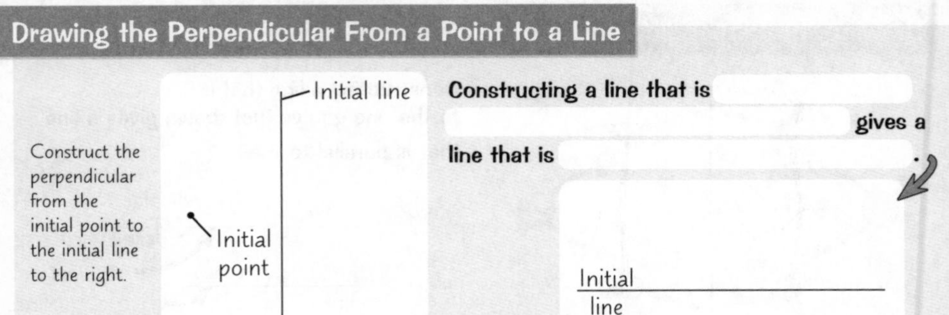

Three Steps to Find Bearings

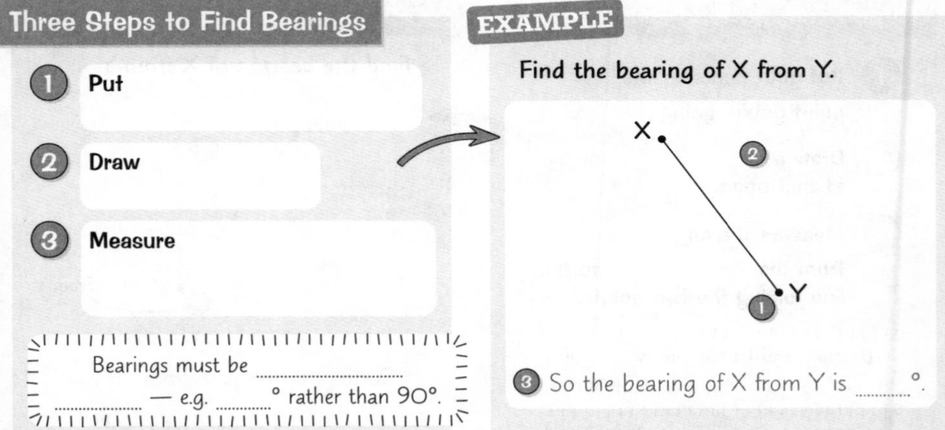

Scale Drawings

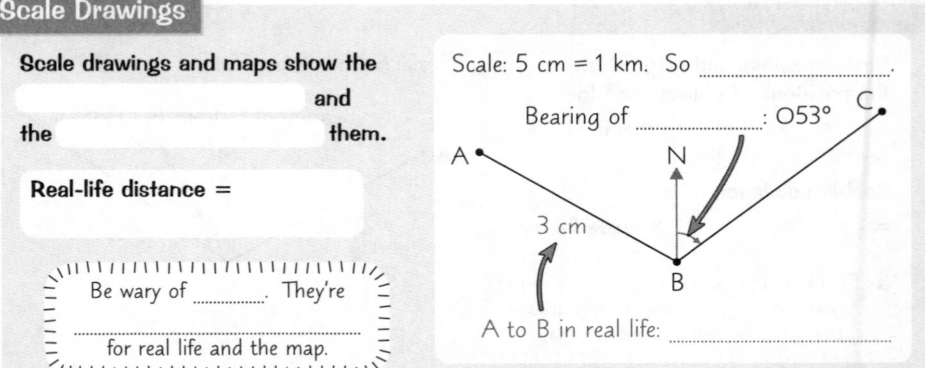

Mixed Practice Quizzes

Oh p.101-110, you were wonderful. I'm so sad to leave you, but it's time. No — it's not you, it's me. At least we've these quizzes to remember our time together.

Quiz 1

Date: ...... / ...... / ......

1) True or false? There are two triangles you could draw if you know two sides and the angle between them.
2) What shape is the locus of points at a fixed distance from a given point?
3) How many vertices, edges and faces does a cube have?
4) What should you do with compass settings when constructing 90° angles?
5) How do you find the volume of a frustum?
6) What do you need to construct to draw a line parallel to an initial line?
7) In which direction do you measure bearings: clockwise or anticlockwise?
8) If a shape is enlarged by a scale factor of n, how many times bigger is the new area than the old area?
9) What is the formula for the volume of a prism?
10) When you're constructing an SAS triangle, what should you do next after marking the angle with a dot?

Total:

Quiz 2

Date: ...... / ...... / ......

1) If a shape is enlarged by a scale factor of 2, what is the ratio of the old volume to the new volume?
2) What does the locus of points equidistant from two lines do to the angle between them?
3) What is the formula for the surface area of a cone?
4) Which type of projection shows the view from above?
5) What does a rate of flow tell you?
6) How many arcs do you need to draw to construct a 60° angle?
7) Point B is due east of point A. What is the bearing of B from A?
8) How do you find the surface area of a prism or pyramid?
9) What is a locus?
10) What is the first thing you need to do when constructing a triangle?

Total:

Mixed Practice Quizzes

Quiz 3

Date: ...... / ...... / ......

1) Which locus is the same as the perpendicular bisector of a line AB?
2) How many vertices, edges and faces does a triangular prism have?
3) What is the final thing you should do when constructing a triangle?
4) What is the volume of a pyramid with base area 6 cm^2 and height 4 cm?
5) How do you find the real-life distance from a scale drawing?
6) What do you use the formula $2\pi rh + 2\pi r^2$ to find?
7) When you're constructing an ASA triangle, what should you do next after marking the angles with dots?
8) At which point do you need to draw a north line to find the bearing of A from B?
9) What is the formula for the volume of a sphere?
10) What should you do with compass settings when constructing 60° angles?

Total:

Quiz 4

Date: ...... / ...... / ......

1) How do you construct the locus of points equidistant from two lines?
2) The scale on a map is 2 cm = 1 km. If A is 10 cm from B on the map, how far is it from B in real life?
3) Which locus has a sausage shape?
4) Which height do you use to find the surface area of a cone?
5) When you're constructing a triangle with three known sides, what do you do after drawing and labelling the base line?
6) What do you use the formula $4\pi r^2$ to find?
7) How would you write 60° as a bearing?
8) When you're constructing the perpendicular from a point to a line, where do you draw your first arcs from?
9) What is the formula for the volume of a cone?
10) How many triangles could you draw if you know two sides and the angle that isn't between them?

Total:

Pythagoras' Theorem

First Go: ...../...../.....

Pythagoras' Theorem

Uses two sides
to find :

+ =

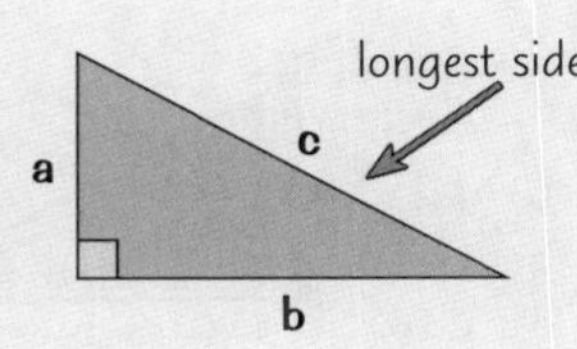

longest side =

Pythagoras' theorem only works for
....
....

Find a Missing Length

1. Write down .
2. Put in .
3. equation.
4. Take .
5. Give answer in .

EXAMPLE

Find the length of AB to 1 d.p.

A, B, C; 9 m; 4 m

1. + =
2. AB^2 + ^{2} = ^{2} c = AC (the longest side)
3. AB^2 = ^{2} – ^{2} = – =
4. AB = = m
5. m (1 d.p.)

Find Distance Between Points

1. Sketch .
2. Subtract
   to find lengths.
3. Use
   to find .
4. Give answer in
   .

The distance is the ...., so you don't need to rearrange the equation.

EXAMPLE

Point L has coordinates (–1, 0).
Point M has coordinates (3, –2).
Find the exact distance LM.

1. L( )

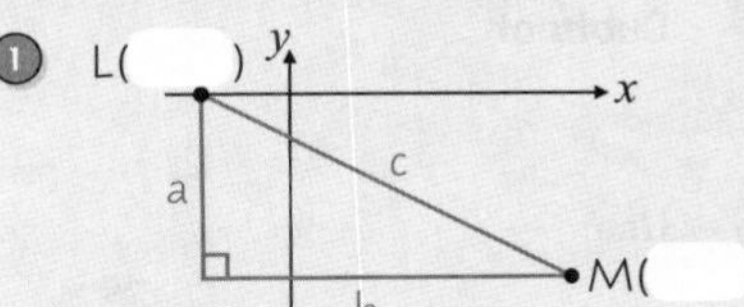

2. Length of side a = – =
   Length of side b = – =
3. $a^2 + b^2 = c^2$
   + = c^2
   c^2 = + =
4. c = =

'Exact' means leave it in .... form (simplified if possible).

Second Go:
...../...../.....

Pythagoras' Theorem

Pythagoras' Theorem

Uses
to find :

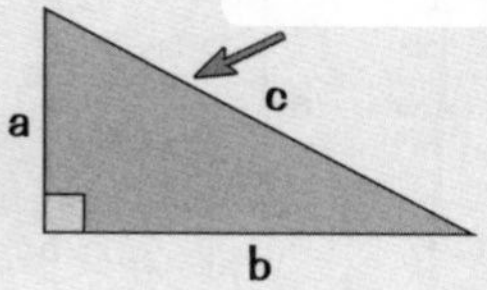

Pythagoras' theorem
....
....
....

Find a Missing Length

1. Write
2.
3.
4.
5. Give answer

EXAMPLE

Find the length of AB to 1 d.p.

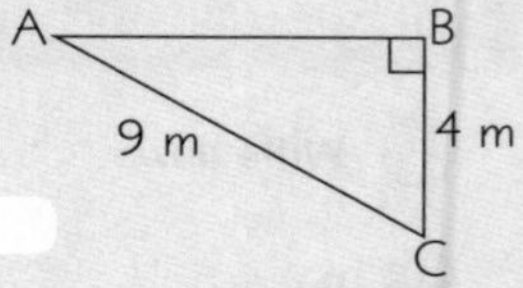

1.
2. ← c = AC (the longest side)
3. $AB^2 =$
4. $AB =$
5.

Find Distance Between Points

1. Sketch triangle.
2. Subtract
3. Use
4. Give answer

The distance is the ...., so you don't need to ....

EXAMPLE

Point L has coordinates (–1, 0).
Point M has coordinates (3, –2).
Find the exact distance LM.

1.

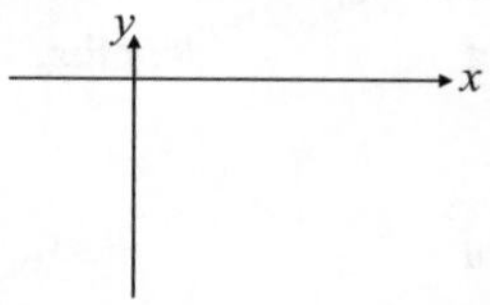

2.
3. $= c^2$
 $= c^2$
 $c^2 =$
4. $c =$

'Exact' means
....
....
(simplified if possible).

Trigonometry

First Go: ..... /..... /.....

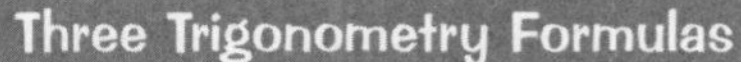

Three Trigonometry Formulas

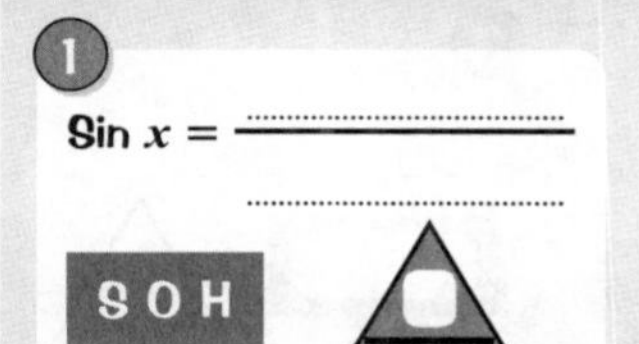

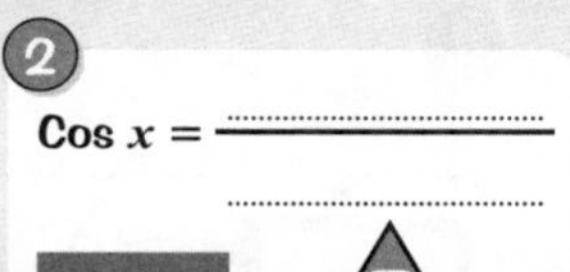

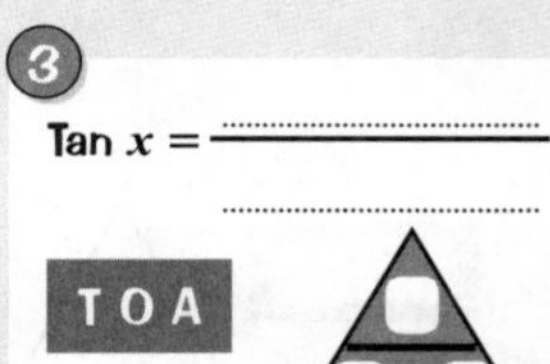

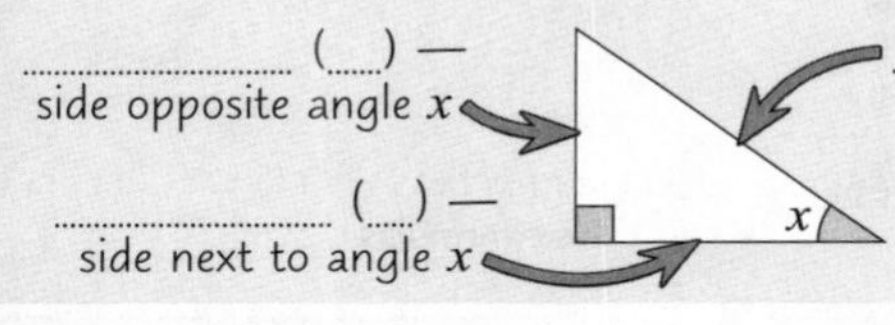

.... (....) — the longest side (opposite right angle)

These formulas only work on ....

Find a Missing Length

1. Label sides ....
2. Choose ....
3. Use a formula triangle to ....
4. Put in .... and work out ....

EXAMPLE

Find the length of g to 2 s.f.

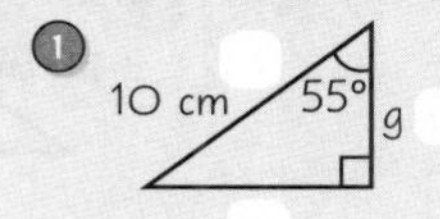

2 .... and .... are involved, so use .... .
SOH CAH TOA

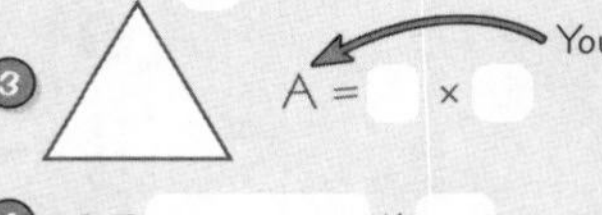

4 g = ... × ....
= .... = ... cm (2 s.f.)

Find a Missing Angle

1. Label sides ....
2. Choose ....
3. Use a formula triangle to ....
4. Put in ....
5. Take .... to find ....

EXAMPLE

Find angle x to 1 d.p.

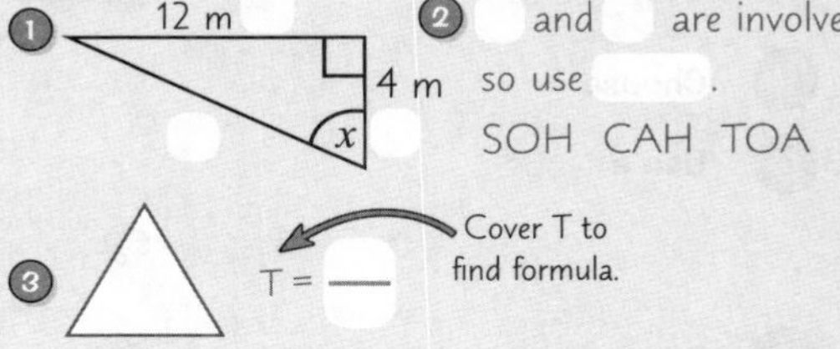

2 .... and .... are involved, so use .... .
SOH CAH TOA

4 tan x = —— = ...

5 x = ... = ...° = ...° (1 d.p.)

Second Go: ...../...../.....

Trigonometry

Three Trigonometry Formulas

1

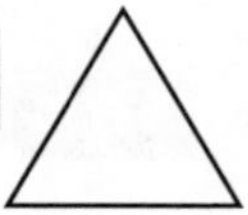

2

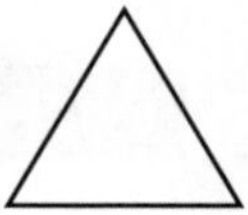

3

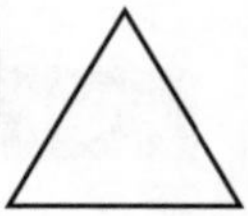

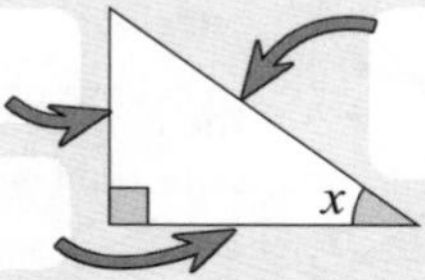

These formulas ...

Find a Missing Length

1. Label
2. Choose
3. Use a
4. Put in

EXAMPLE

Find the length of g to 2 s.f.

① 10 cm, 55°, g

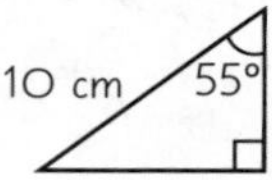

②

SOH CAH TOA

③ You're finding A.

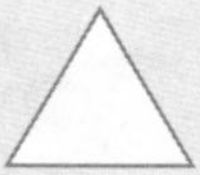

A =

④ g =

=

Find a Missing Angle

1. Label
2. Choose
3. Use a
4. Put in
5. Take

EXAMPLE

Find angle x to 1 d.p.

① 12 m, 4 m, x

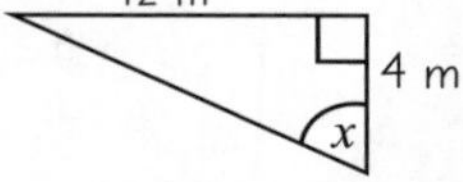

②

SOH CAH TOA

③ Cover T to find formula.

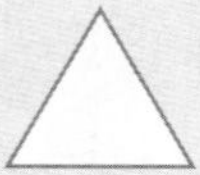

T =

④ tan x =

⑤ x =

=

Common Trig Values and the Sine Rule

First Go: ..... /..... /.....

Common Trig Values

These values can be worked out from these two triangles.

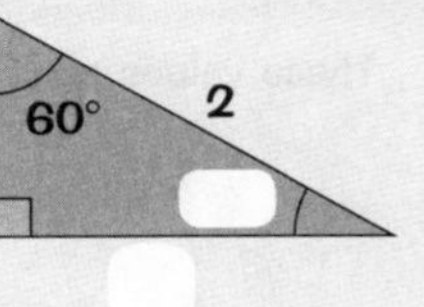

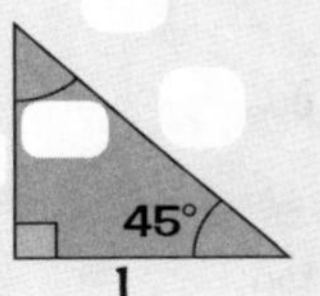

sin $\quad = \frac{1}{2}$ sin 60° = sin $\quad = \frac{1}{\sqrt{2}}$

cos 30° = cos $\quad = \frac{1}{2}$ cos 45° =

tan $\quad = \frac{1}{\sqrt{3}}$ tan 60° = tan $\quad = 1$

These values cannot be worked out using triangles.

sin 0° = sin $\quad$ = 1

cos 0° = cos $\quad$ = 0 tan 0° =

Use common trig values to find .... in triangles.

The Sine Rule

$$\frac{\quad}{\sin A} = \frac{b}{\quad} = \frac{c}{\sin \quad}$$

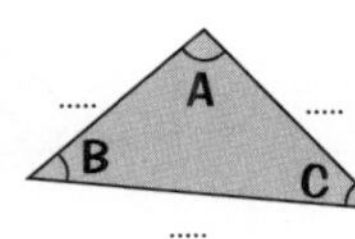

You only use .... of the formula at a time. You can turn the formula .... if you're finding an angle.

Use when given:

.... + **ANY SIDE**

.... + **ANGLE NOT ENCLOSED BY THEM**

EXAMPLE

Find the length AC.

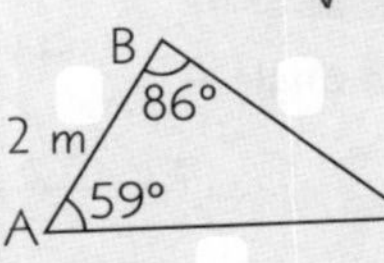

1. Find missing angle.
   C = $\quad - \quad - \quad = $
2. Put numbers in sine rule.
   $\quad = \quad \Rightarrow \frac{AC}{\sin} = \frac{\quad}{\sin}$
3. Rearrange to find length.
   AC = $\frac{\times}{\sin}$ = m (2 s.f.)

EXAMPLE

Find angle A.

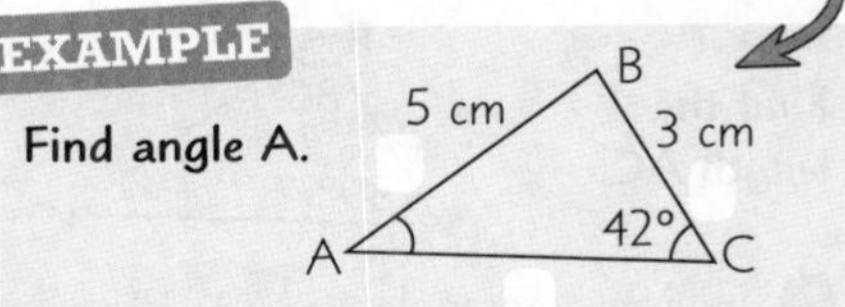

1. Put numbers in sine rule.
   $\quad = \quad \Rightarrow \frac{\quad}{\sin A} = \frac{\quad}{\sin}$
2. Rearrange to find sin A.
   sin A = $\frac{\times}{\quad}$ =
3. Take inverse to find angle.
   A = = ° (1 d.p.)

Second Go: ...../...../.....

Common Trig Values and the Sine Rule

Common Trig Values

These values can be worked out from these two triangles.

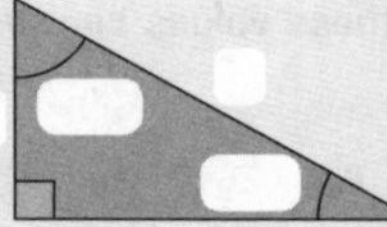

sin = sin = sin =

cos = cos = cos =

tan = tan = tan =

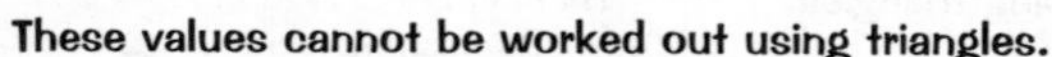

These values cannot be worked out using triangles.

sin = sin =

cos = cos = tan =

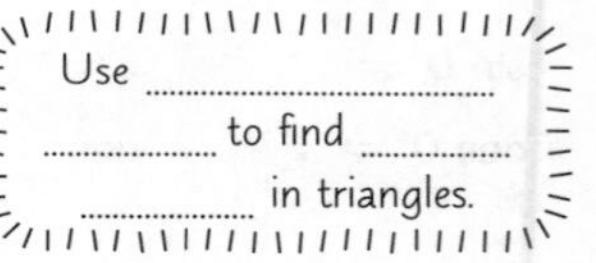

Use .... to find .... in triangles.

The Sine Rule

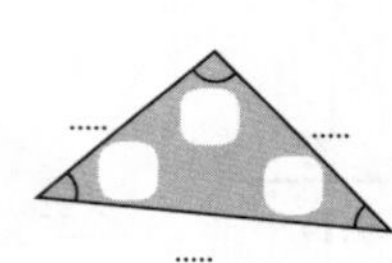

You only use .... of the formula at a time. You can turn the formula .... if you're finding an ....

Use when given:

.... + ...

.... + ...

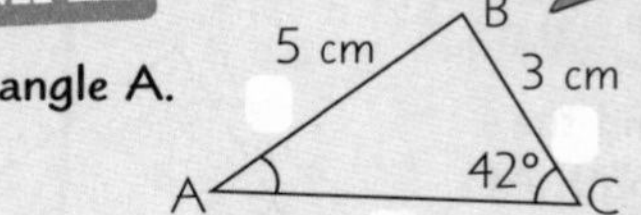

EXAMPLE

Find the length AC.

1. Find missing angle.
2. Put numbers in sine rule. ⇒
3. Rearrange to find length.

EXAMPLE

Find angle A.

1. Put numbers in sine rule. ⇒
2. Rearrange to find sin A.
3. Take inverse to find angle.

The Cosine Rule and Area of a Triangle

First Go: ..... /..... /.....

The Cosine Rule

To find ....: $a^2 = \quad + \quad - 2 \quad \cos$

To find ....: $\cos A = \frac{\quad + \quad - \quad}{\quad}$

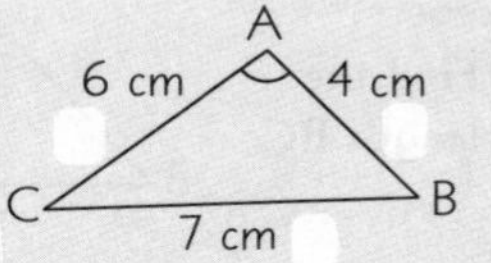

Use when given:

.... + ANGLE ENCLOSED BY THEM

...................................., NO ANGLES

EXAMPLE

Find the length BC.

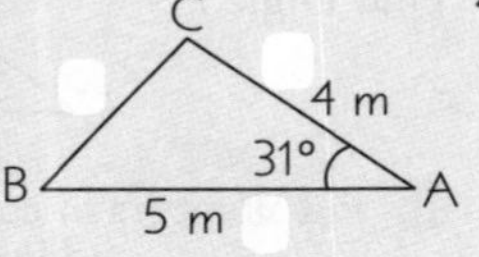

1. Put numbers in cosine rule.

$a^2 = \quad + \quad - 2 \quad \cos$

$= \quad + \quad - 2 \times \quad \times \quad \cos$

$=$

2. Take square root to find length.

a = = m (2 s.f.)

EXAMPLE

Find angle A.

1. Put numbers in cosine rule.

$\cos A = \frac{\quad + \quad - \quad}{\quad} = \frac{\quad + \quad - \quad}{\quad}$

$=$

2. Take inverse to find angle.

A = = ° (1 d.p.)

Area of Triangle

Area of triangle = $\frac{1}{2}$ .... sin .....

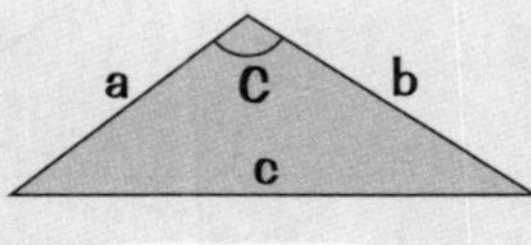

Use when given and the angle by them.

Two steps to find the area of a triangle:

1. Label the and .
2. Put in .

EXAMPLE

Find the area of triangle PQR.

1.

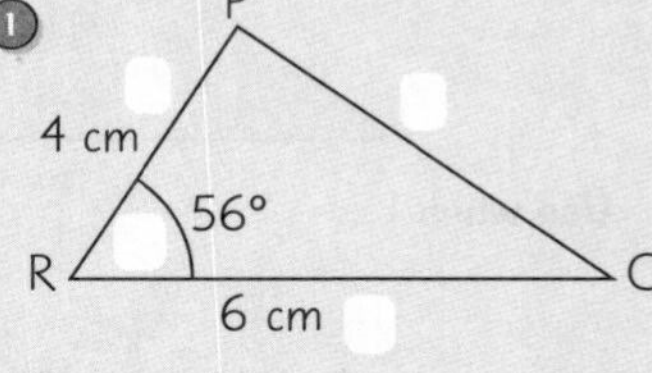

2. Area = $\frac{1}{2}$ sin

$= \frac{1}{2} \times \quad \times \quad \times \sin$

= cm² (2 s.f.)

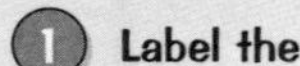

Second Go:
...... /...... /......

The Cosine Rule and Area of a Triangle

The Cosine Rule

To find a side:

To find an angle:

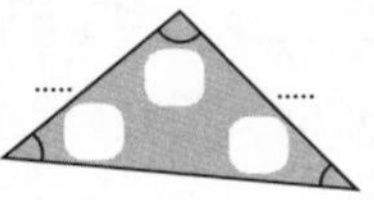

Use when given:

.... + ...

....

...., ...

EXAMPLE

Find the length BC.

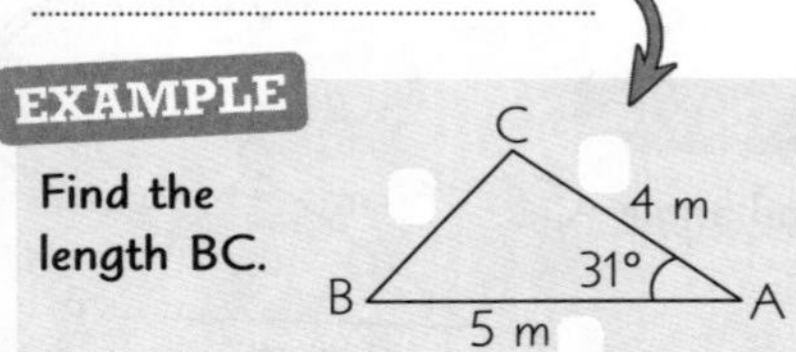

1 Put numbers in cosine rule.

$a^2 =$

$=$

$=$

2 Take square root to find length.

$a =$

EXAMPLE

Find angle A.

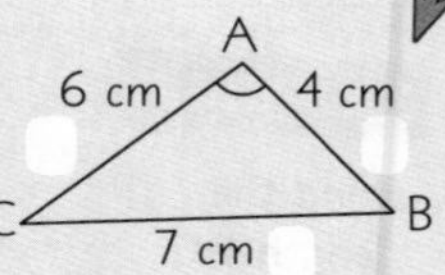

1 Put numbers in cosine rule.

$\cos A =$

2 Take inverse to find angle.

$A =$

$=$

Area of Triangle

Area of triangle =

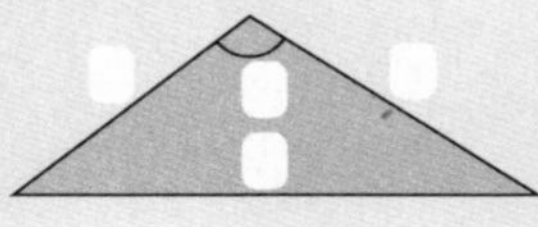

Use when

Two steps to find the area of a triangle:

1

2

EXAMPLE

Find the area of triangle PQR.

1

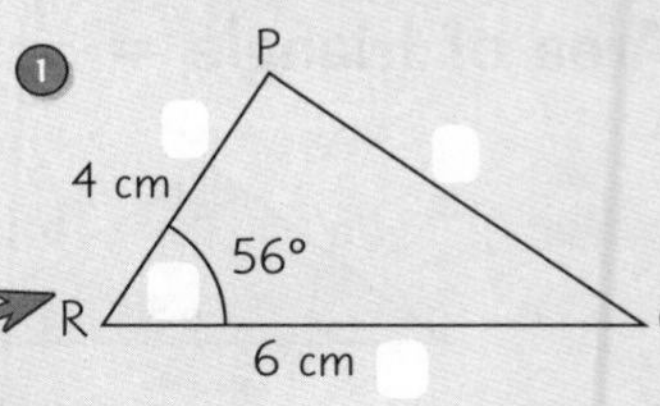

2 Area =

=

=

3D Pythagoras and Trigonometry

First Go: ..... /..... /.....

3D Pythagoras

To find the length of a ______ of a cuboid:

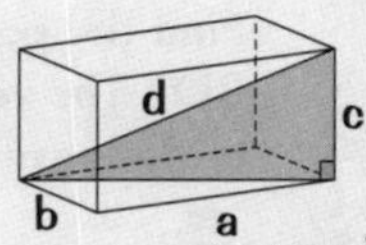

Two steps for other 3D shapes:

1. Form a ______ that has ______ within the 3D shape.
2. Put numbers into ______.

EXAMPLE

Find the exact length of BD. The vertical height of the prism is 4 m.

1
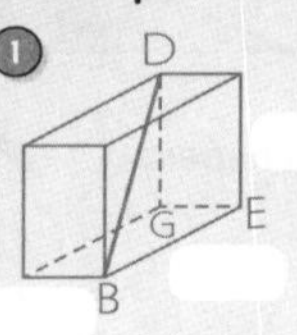

2 ___ + ___ + ___ = d^2

___ + ___ + ___ = BD^2

BD = ___ = ___ m

Angle Between Line and Plane

1. Draw ______ between the ______ and ______.
2. Sketch ______ in 2D.
3. Use ______ to find any ______.
   You need to know two sides.
4. Use ______ to find ______.

EXAMPLE

Find the angle that the diagonal BH makes with the cuboid's base.

1
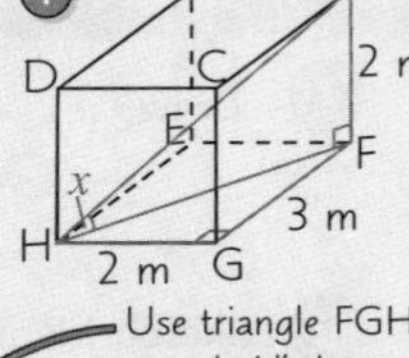

2
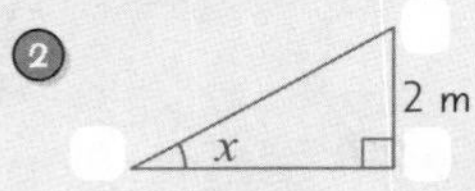

Use triangle FGH on cuboid's base.

3 FH^2 = ___ + ___ = ___, so FH = ___

4 T = ___ ⇒ tan x = ___ = ___

⇒ x = ___

= ___ ° (1 d.p.)

Find a Length Using an Angle

1. Draw ______ containing ______ and ______.
2. Sketch ______ in 2D.
3. Use ______ to find a ______.
4. Use ______ to find ______.

EXAMPLE

In this square-based pyramid, angle ACE = 50°. Find length AC.

1
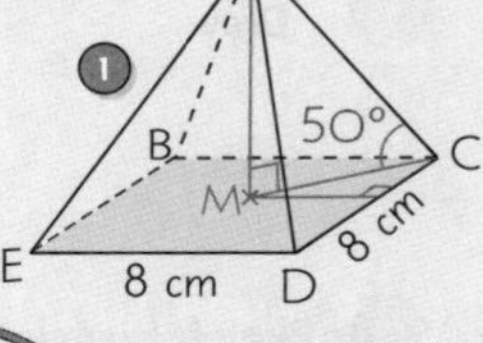

2
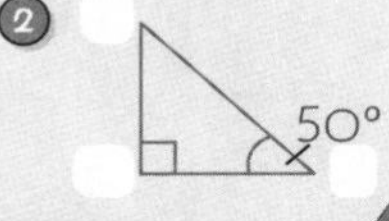

Use triangle on base with shorter sides of length 4 cm (as M is the midpoint).

3 MC^2 = ___ + ___ = ___

so MC = ___ cm

4 C = ___ ⇒ cos 50° = ___

AC = ___ ÷ ___ = ___ cm (2 s.f.)

3D Pythagoras and Trigonometry

3D Pythagoras

To find the length of a :

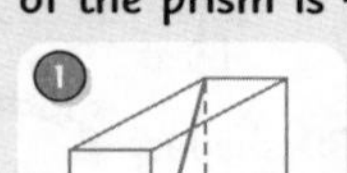

Two steps for other 3D shapes:

1. Form
2. Put

EXAMPLE

Find the exact length of BD. The vertical height of the prism is 4 m.

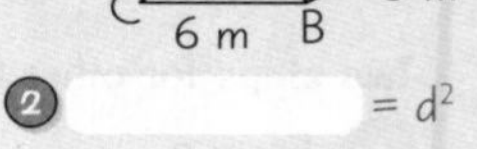

1

2 $= d^2$

$= BD^2$

BD =

=

Angle Between Line and Plane

1. Draw
2. Sketch triangle in 2D.
3. Use

   You need to know two sides.
4. Use

EXAMPLE

Find the angle that the diagonal BH makes with the cuboid's base.

1

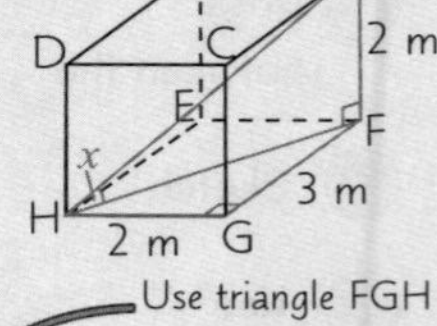

2

Use triangle FGH on cuboid's base.

3 $FH^2 =$, so FH =

4 T = $\Rightarrow$ $\tan x =$

$\Rightarrow x =$

Find a Length Using an Angle

1. Draw
2. Sketch triangle in 2D.
3. Use
4. Use

EXAMPLE

In this square-based pyramid, angle ACE = 50°.
Find length AC.

1

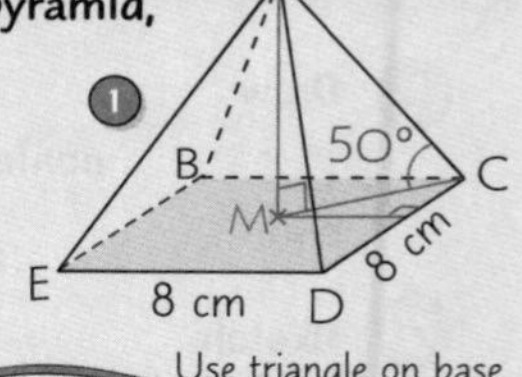

2

Use triangle on base with shorter sides of length 4 cm (as M is the midpoint).

3 $MC^2 =$,

so MC =

4 C = $\Rightarrow$ =

AC =

Vectors

First Go: ..... /..... /.....

Vector Notation and Ratios

This vector can be written as , **b**, or $\begin{pmatrix}4\\-1\end{pmatrix}$.

Ratios can show of sections on a line.

If you know one vector, you can use it to find others.

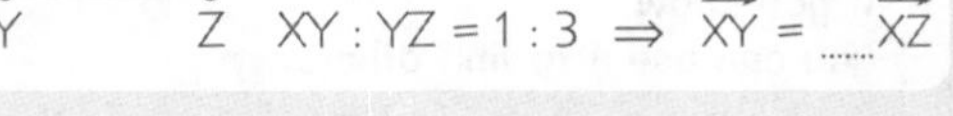

Multiplying a Vector by a Scalar

Scalar multiples are ....

Multiplying a vector by:

+ a positive number changes its only.

− a negative number the direction too.

c
1.5c
−0.5c
−2c

Adding and Subtracting Vectors

To describe a movement between points:

1. Find made up of .
2. vectors along route. vectors travelled in direction.

For column vectors: add/subtract , then .

E.g. $\begin{pmatrix}4\\-1\end{pmatrix} - \begin{pmatrix}2\\3\end{pmatrix} =$ ....

EXAMPLE

M is the midpoint of AB. Find vector $\overrightarrow{CA}$.

C, c, A, M, B, b

1. $\overrightarrow{AM} =$ as M is the midpoint. So $\overrightarrow{CA} =$ + + .
2. = =

You're going backwards along b, so subtract.

Showing Points are on a Straight Line

Points A, B, C lie on a straight line if $\overrightarrow{AB}$ is a of $\overrightarrow{BC}$ or $\overrightarrow{AC}$.

1. Work out the between .
2. Check vectors are of each other.
3. your reasoning.

EXAMPLE

Show that ABC is a straight line.

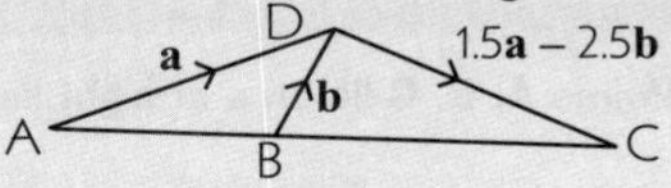

1. Find $\overrightarrow{AB}$ and $\overrightarrow{BC}$.
   $\overrightarrow{AB} =$
   $\overrightarrow{BC} =$
   $=$
2. $\overrightarrow{BC} =$, so $\overrightarrow{BC} =$
3. $\overrightarrow{BC}$ is a of $\overrightarrow{AB}$, so ABC is a .

Second Go:
...../...../.....

Vectors

Vector Notation and Ratios

This vector can be written as , **b**, or .

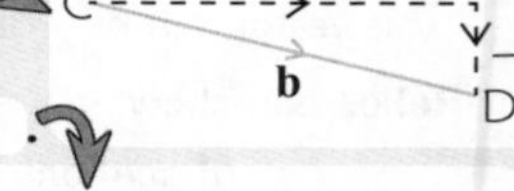

Ratios can show of sections .

If you know ,
you can use it to find others.

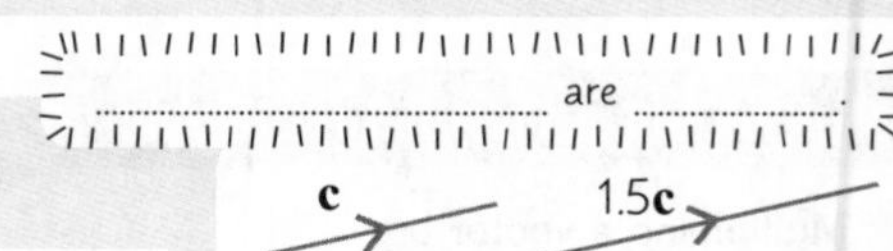

Multiplying a Vector by a Scalar

.... are ....

Multiplying a vector by:

+ a positive number .

− a negative number
.

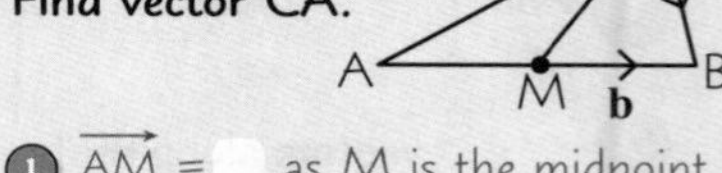

Adding and Subtracting Vectors

To describe a movement between points:

1. Find
2. Add
   Subtract

For column vectors:

E.g. $\begin{pmatrix}4\\-1\end{pmatrix} - \begin{pmatrix}2\\3\end{pmatrix} =$ ....

EXAMPLE

M is the midpoint of AB.
Find vector $\overrightarrow{CA}$.

A M **b** B C **c**

1. $\overrightarrow{AM}$ = as M is the midpoint.
   So
2.

You're going backwards along b, so subtract.

Showing Points are on a Straight Line

Points A, B, C lie on a straight line if

1. Work out
2. Check
3. Explain

EXAMPLE

Show that ABC is a straight line.

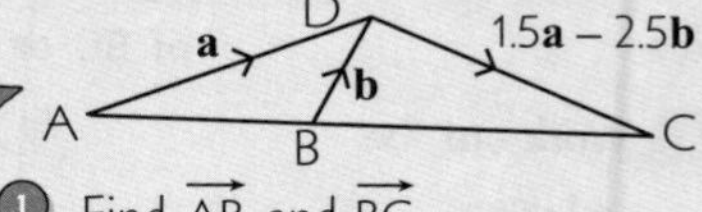

1. Find $\overrightarrow{AB}$ and $\overrightarrow{BC}$.
2. , so
3.

Mixed Practice Quizzes

Onwards to vector-y... But first, a few quizzes to check everything from p.113-124 is firmly embedded in your brain. Mark them yourself and jot down your score.

Quiz 1

Date: ...... / ...... / ......

1) What is the name given to the longest side of a right-angled triangle?
2) What happens to a vector when you multiply it by a positive number?
3) What is the exact value of cos 45°?
4) Which trig formula should you use if you know the opposite and adjacent?
5) Give the formula for the sine rule.
6) When might you need to use Pythagoras in a 3D shape, given an angle?
7) When using trig, what is the first thing you should do to find a missing side or angle?
8) Find the area of an equilateral triangle with sides of length 2 cm.
9) A right-angled triangle has shorter sides of 3 cm and 4 cm. What is the length of the longest side?
10) How can you show that points P, Q and R lie on a straight line?

Total:

Quiz 2

Date: ...... / ...... / ......

1) What does SOH CAH TOA stand for?
2) In a column vector, what do the top and bottom numbers show?
3) What is the trig formula for the area of a triangle?
4) What is the formula for Pythagoras' theorem?
5) Which trig value is equal to $\sqrt{3}$?
6) Which trig formula should you use if you know the opposite and hypotenuse?
7) When using Pythagoras' theorem, what is the next step once you've put in the numbers and rearranged the equation?
8) If you know two angles and a side, should you use the sine or cosine rule?
9) PQ : QR = 2 : 5. $\overrightarrow{PR} = \underline{a}$. Give $\overrightarrow{QR}$ in terms of $\underline{a}$.
10) A cuboid has sides of length 3 cm, 5 cm and 7 cm. Find the exact length of its longest diagonal.

Total:

Mixed Practice Quizzes

Quiz 3

Date: ...... / ...... / ......

1) What is the value of sin 30°?
2) If you know all 3 sides and no angles, should you use the sine or cosine rule?
3) For a right-angled triangle with sides a, b and c, rearrange Pythagoras' theorem to find a shorter side, a.
4) When using trig, what is the next thing you have to do to find an angle once you've put the numbers into the formula?
5) Which formula would you use to find the length of the longest diagonal in a cuboid?
6) True or false? Scalar multiples of vectors are perpendicular.
7) What is the formula triangle for cos?
8) How do you use Pythagoras' theorem to find the distance between 2 points?
9) What is the cosine rule for finding an angle?
10) In triangle ABC, $\overrightarrow{AB} = \underline{a}$ and $\overrightarrow{CB} = \underline{b}$. Find $\overrightarrow{AC}$ in terms of $\underline{a}$ and $\underline{b}$.

Total:

Quiz 4

Date: ...... / ...... / ......

1) A right-angled triangle has a hypotenuse of 13 cm and another side of 5 cm. What is the length of the third side?
2) What happens to a vector when you multiply it by a negative number?
3) What is the value of cos 0°?
4) True or false? The sine and cosine rules work on any triangles.
5) What is the cosine rule for finding a side?
6) What are the side lengths of the right-angled triangle used to work out the exact values of sin, cos and tan of 60° and 30°?
7) Which trig formula should you use if you know the adjacent and hypotenuse?
8) What is the first thing you should do to find the angle between a line and a plane in a 3D shape?
9) Find $\begin{pmatrix} -2 \\ 3 \end{pmatrix} + \begin{pmatrix} 4 \\ 5 \end{pmatrix}$.
10) What information do you need to find the area of a triangle using trig?

Total:

Probability Basics

First Go: ..... /..... /.....

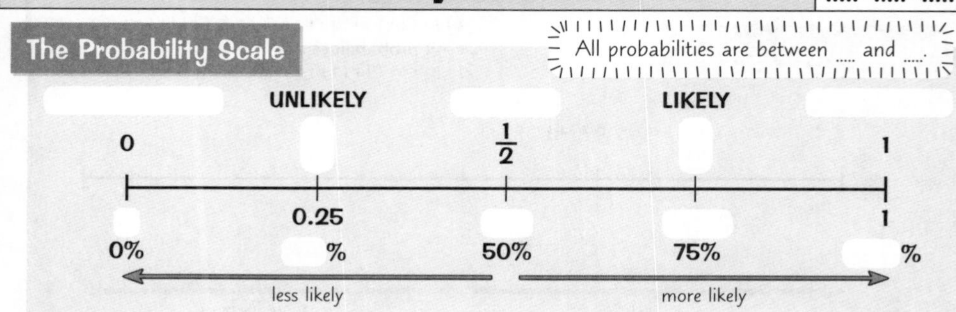

The Probability Scale

All probabilities are between ..... and .....

	UNLIKELY		LIKELY	
0		$\frac{1}{2}$		1
	0.25			1
0%	%	50%	75%	%

← less likely | more likely →

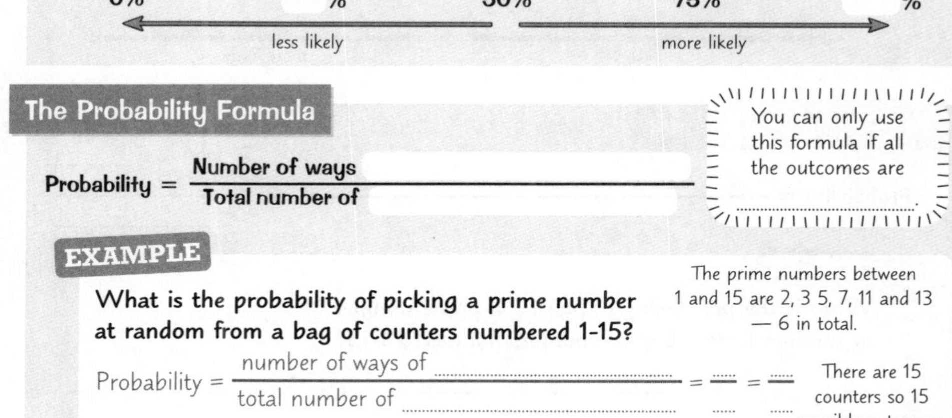

The Probability Formula

$$\text{Probability} = \frac{\text{Number of ways}}{\text{Total number of}}$$

You can only use this formula if all the outcomes are ....

EXAMPLE

What is the probability of picking a prime number at random from a bag of counters numbered 1-15?

The prime numbers between 1 and 15 are 2, 3 5, 7, 11 and 13 — 6 in total.

Probability = $\frac{\text{number of ways of ....}}{\text{total number of ....}}$ = $\frac{\text{.....}}{\text{.....}}$ = $\frac{\text{.....}}{\text{.....}}$

There are 15 counters so 15 possible outcomes.

Probabilities of Events

If only one possible outcome can happen at a time, the .... of all possible outcomes add up to ..... As events either happen or .... :

P(............) + P(event doesn't happen) = ...

So:

P(event doesn't happen) = ... – P(............)

EXAMPLE

The probability of getting a 5 on a spinner is 0.65. What is the probability of not getting a 5?

P(not 5) = ... – P(5)

= ..... – 0.65

= .....

Sample Space Diagrams

These show ....

E.g. All possible outcomes when two spinners numbered 1, 2, 3 and 2, 4, 6 are spun and the results ....

×	1	2	3
2	2		6
4		8	
6	6		

The Product Rule

Number of ways to carry out a combination of activities = number of ways to carry out each activity .... together

Number of ways to roll 3 fair 6-sided dice = ... × ..... × ..... = ...

Second Go:
...../...../.....

Probability Basics

The Probability Scale

All probabilities are ....

less likely — more likely

The Probability Formula

Probability = ______________________

You can only use this formula if ....

EXAMPLE

What is the probability of picking a prime number at random from a bag of counters numbered 1-15?

Probability = ... = ... = ...

Probabilities of Events

If only one possible outcome can happen at a time, the probabilities of all possible outcomes ....

As events either .... or .... :

So:

EXAMPLE

The probability of getting a 5 on a spinner is 0.65. What is the probability of not getting a 5?

P(not 5) = ...
= ....
= ....

Sample Space Diagrams

×			

E.g. All possible outcomes when two spinners numbered 1, 2, 3 and 2, 4, 6 are spun and the results ....

The Product Rule

Number of ways to carry out a .... of activities = number of ways to carry out each activity ....

Number of ways to roll 3 fair 6-sided dice = ... = ...

Probability Experiments

First Go: ..... / ..... / .....

Repeating Experiments

FAIR — every outcome is ______ to happen.

BIASED — some outcomes are ______ than others.

Relative frequency = $\frac{\text{\_\_\_\_\_\_}}{\text{Number of times \_\_\_\_\_\_}}$

Use relative frequencies to ______ probabilities.

The more times you do an experiment, the more ______ the estimate is likely to be.

EXAMPLE

A spinner labelled A to D is spun 100 times. It lands on C 48 times. Find the relative frequency of spinning a C and say whether you think this spinner is biased.

Relative frequency of C = $\frac{\dots}{\dots}$ = ...

If the spinner was fair, you'd expect the relative frequency of C to be ..... ÷ ..... = ... is much larger than ...., so the spinner is probably ....

Frequency Trees

Used to record results when experiments have ______ ______. For example:

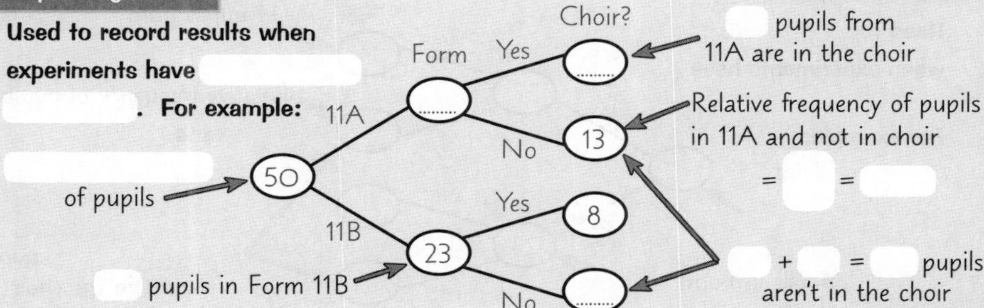

Expected Frequency

EXPECTED FREQUENCY — how many times you'd ______ something to happen in a certain ______.

Expected frequency
= ______ × ______

Use the .... from previous experiments if you don't know the probability.

EXAMPLE

A fair 6-sided dice is rolled 360 times. How many times would you expect it to land on 4?

P(4) = ______

Expected frequency of 4
= ______ × ______ = ______

Second Go: ...../...../.....

Probability Experiments

Repeating Experiments

FAIR —

BIASED —

Relative frequency = $\frac{\quad}{\quad}$

Use relative frequencies to . The you do an experiment, the is likely to be.

EXAMPLE

A spinner labelled A to D is spun 100 times. It lands on C 48 times. Find the relative frequency of spinning a C and say whether you think this spinner is biased.

Relative frequency of C = $\frac{\dots}{\dots}$ = ...

If the spinner was ...., you'd expect the relative frequency of C to be .... is much .... than ...., so the spinner is probably ....

Frequency Trees

Used to when experiments have .

For example:

Choir?

Form

11A — 27: Yes — 14; No — ......

11B — 23: Yes — ......; No — 15

Total: ......

pupils in Form

14 pupils from

Relative frequency of pupils in and = =

pupils aren't in the choir

Expected Frequency

EXPECTED FREQUENCY —

Expected frequency =

Use the .... from .... if you don't know the ....

EXAMPLE

A fair 6-sided dice is rolled 360 times. How many times would you expect it to land on 4?

P(4) =

Expected frequency of 4

= =

The AND/OR Rules

First Go: ..... /..... /.....

The AND Rule for Independent Events

INDEPENDENT EVENTS — where one event happening ______ the ______ of another event happening.

If you select a second item after ______ the first, the events are ______.

For independent events A and B:

P(A and B) = ______ × ______

EXAMPLE

A fair dice is rolled and a fair coin is tossed. What is the probability of rolling a 2 and getting heads?

P(2) = ___ and P(heads) = ___

Rolling a dice and tossing a coin are independent, so:

P(2 and heads) = ___ × ___ = ___

The OR Rule

For any events A and B:

P(A or B) = ______ + ______ − ______

If A and B are mutually exclusive, P(A and B) = ___. So the OR rule becomes:

P(A or B) = ______ + ______

Mutually exclusive events ....

EXAMPLE

A fair dice is rolled and a fair coin is tossed. What is the probability of rolling a 2 or getting heads?

From above, P(2) = ___ P(heads) = ___

and P(2 and heads) = ___

So P(2 or heads) = ___ + ___ − ___

= ___

Conditional Probability

DEPENDENT EVENTS — where one event happening ______ the ______ of another event happening.

If you select a second item without ______ the first, the events are ______.

CONDITIONAL PROBABILITY OF A GIVEN B — the probability of event A happening ______ event B happens.

For dependent events A and B, the AND rule is:

P(A and B) = ______ × ______

P(B given A) can be written ______.

EXAMPLE

The probability that Abi has pasta for tea is 0.6. The probability that Abi has yoghurt for pudding given that she has pasta for tea is 0.7. What is the probability that Abi has pasta for tea and yoghurt for pudding?

P(pasta and yoghurt)

= ______ × ______

= ___ × ___ = ___

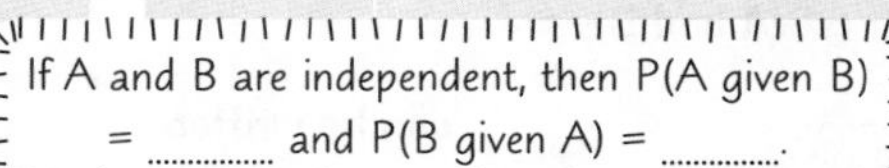

If A and B are independent, then P(A given B) = ... and P(B given A) = ...

Second Go: ...../...../.....

The AND/OR Rules

The AND Rule for Independent Events

INDEPENDENT EVENTS —

If you select a second item , the events are independent.

For independent events A and B:

P(A and B) =

EXAMPLE

A fair dice is rolled and a fair coin is tossed. What is the probability of rolling a 2 and getting heads?

P( ) = and P( ) =

Rolling a dice and tossing a coin are , so:

P( ) = =

The OR Rule

For any events A and B:

P(A or B) =

If A and B are ,

P( ) = 0. So the OR rule becomes:

P(A or B) =

.... can't happen together.

EXAMPLE

A fair dice is rolled and a fair coin is tossed. What is the probability of rolling a 2 or getting heads?

From above, P( ) = ,

P( ) = and

P( ) =

So P( )

= =

Conditional Probability

DEPENDENT EVENTS —

If you select a second item , the events are dependent.

CONDITIONAL PROBABILITY OF A GIVEN B —

For dependent events A and B, the AND rule is:

P(A and B) =

can be written .

EXAMPLE

The probability that Abi has pasta for tea is 0.6. The probability that Abi has yoghurt for pudding given that she has pasta for tea is 0.7. What is the probability that Abi has pasta for tea and yoghurt for pudding?

P( )

=

= =

If A and B are independent, then .... and ....

Tree and Venn Diagrams

First Go: ..... /..... /.....

Tree Diagrams

Used to work out probabilities for ___ — e.g. for a bag containing 3 red and 2 blue counters that are selected at random without replacement:

Probabilities on each set of branches that meet at a point ___.

First counter — Second counter

$\frac{3}{5}$ R → $\frac{2}{4}$ R: $\frac{3}{5} \times \frac{2}{4} =$ ___

R → ___ B: $\frac{3}{5} \times$ ___ $=$ ___

___ B → $\frac{3}{4}$ R: ___ $\times \frac{3}{4} =$ ___

B → ___ B: ___ $\times$ ___ $=$ ___

___ along the branches to get the end probabilities.

The end probabilities ___.

The events are ___, so the probabilities on the second set of branches ___ depending on the outcomes of the first event.

> If the counters were replaced, the events would be .... so the probabilities on each set of branches would be ....

___ the end probabilities to answer questions:

E.g. P(one red, one blue) = P(___) + P(___)

= ___ + ___ = ___ = ___

Sets and Venn Diagrams

SET — a collection of ___ (e.g. numbers), written in ___.

VENN DIAGRAM — a diagram where ___ are represented by ___.

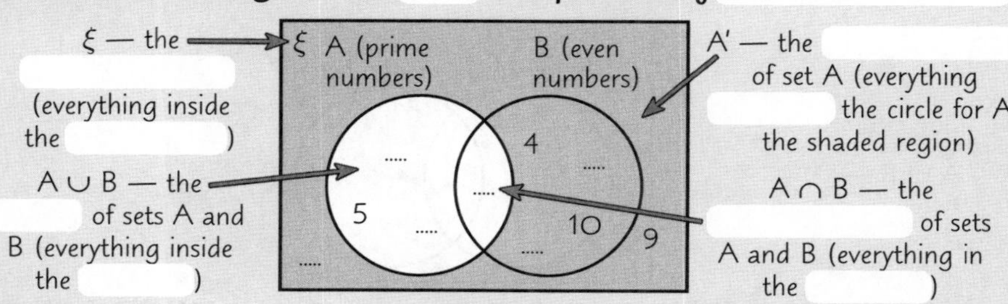

n(A) — the ___ in set A. So here, n(A) = ___ and n(A ∪ B) = ___.

Venn diagrams can show the number of elements instead of the ___.

Use Venn diagrams to find probabilities: E.g. $P(A \cup B) = \frac{n(A \cup B)}{n(\xi)} =$ ___ $=$ ___

Second Go:
...... /...... /......

Tree and Venn Diagrams

Tree Diagrams

Used to work out — e.g. for a bag containing 3 red and 2 blue counters that are selected at random without replacement:

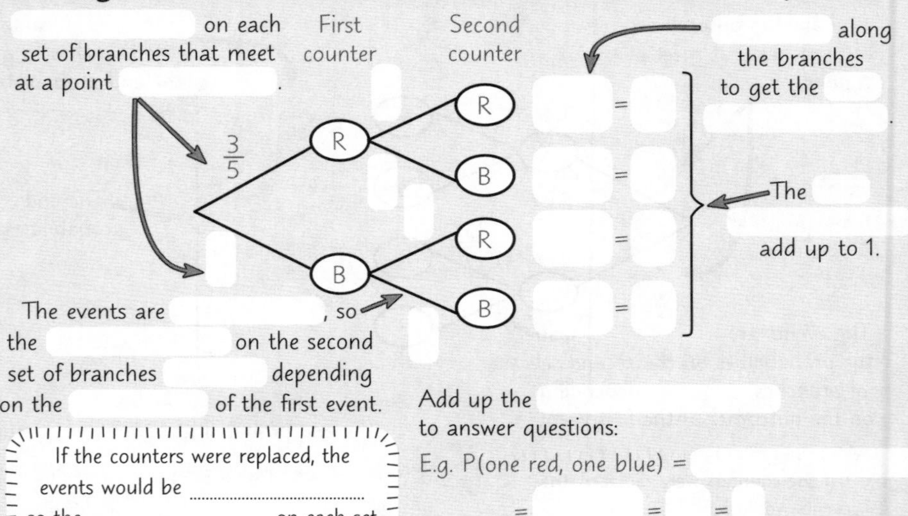

If the counters were replaced, the events would be .... so the .... on each set of branches would be ....

Add up the to answer questions:

E.g. P(one red, one blue) =

= = =

Sets and Venn Diagrams

SET —

VENN DIAGRAM —

ξ —

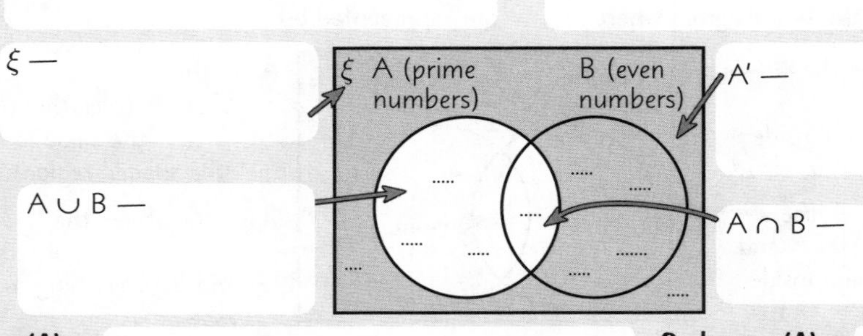

A′ —

A ∪ B —

A ∩ B —

n(A) — . So here, n(A) = and

n(A ∪ B) = . Venn diagrams can show the instead of the .

Use Venn diagrams to find : E.g. P(A ∪ B) = n/n = =

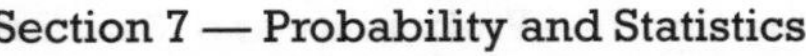

Mixed Practice Quizzes

You knew the probability of facing more quizzes was very likely. It's certain that these will cover p.127-134. Mark them yourself — and make sure you're not biased.

Quiz 1

Date: ...... / ...... / ......

1) How likely is an event with a probability of 0?
2) Find P(A and B) for independent events where P(A) = 0.4 and P(B) = 0.2.
3) What is the difference between independent and dependent events?
4) What number goes at the start of a frequency tree?
5) What is the OR rule for mutually exclusive events A and B?
6) What is the formula for expected frequency?
7) How do you find the end probabilities on a tree diagram?
8) What is meant by the conditional probability of A given B?
9) On a Venn diagram, which area shows $A \cup B$?
10) The probability of picking a red counter from a bag is 0.45. What is the probability of picking a counter that isn't red?

Total:

Quiz 2

Date: ...... / ...... / ......

1) What is a sample space diagram?
2) What is a set?
3) Give the formula used to work out the probability if all possible outcomes are equally likely.
4) What do the end probabilities on a tree diagram add up to?
5) What does the overlap between the circles on a Venn diagram represent?
6) The probability of getting a 1 on a spinner is 0.25. How many times would you expect to get a 1 if you spun the spinner 60 times?
7) How can you improve the accuracy of your relative frequencies?
8) What does it mean if a dice or spinner is fair?
9) How many possible outcomes are there when an 8-sided spinner numbered 1-8 is spun and a coin is tossed?
10) What is the AND rule for dependent events A and B?

Total:

Mixed Practice Quizzes

Quiz 3

Date: ...... / ...... / ......

1) What do the probabilities on a set of branches that meet at a point on a tree diagram add up to?
2) What is the OR rule for any events A and B?
3) How likely is an event with a probability of 1?
4) What is the formula for relative frequency?
5) What is another way of writing P(A given B)?
6) What is the probability of spinning an even number on a fair spinner numbered 1-9?
7) Events A and B are mutually exclusive. P(A) = 0.35 and P(B) = 0.1. Find P(A or B).
8) What are frequency trees used for?
9) On a tree diagram showing dependent events, do the probabilities on each set of branches change or stay the same?
10) Three coins are tossed. How many possible outcomes are there?

Total:

Quiz 4

Date: ...... / ...... / ......

1) What does the notation n(A) represent?
2) What is the AND rule for independent events A and B?
3) What is expected frequency?
4) On a Venn diagram, which area shows $A \cap B$?
5) What word describes an event with a 75% chance of happening?
6) If you know the probability that an event happens, how do you find the probability that the event doesn't happen?
7) What does it mean if a dice or spinner is biased?
8) A dice is rolled fifty times and lands on 6 twelve times. What is the relative frequency of rolling a 6?
9) Are events dependent or independent for selection without replacement?
10) What is the universal set on a Venn diagram?

Total:

Sampling and Data Collection

First Go: ..... /..... /.....

Definitions of Sampling Terms

POPULATION	The you want to find out about.
	A smaller group taken from the population.
RANDOM SAMPLE	Every member of the population has an of being in the sample.
REPRESENTATIVE	the whole population.
	Doesn't fairly represent the whole population.
	Data described by words (not numbers).
QUANTITATIVE DATA	Data described by .
DISCRETE DATA	Data that can only take .
	Data that can take any value in a range.

Choosing a Simple Random Sample

1. Give each member of the population a .
2. Make a list of .
3. the members of the with those numbers.

Random numbers can be chosen using a ...., or from a ....

Spotting Bias

Two things to think about:

1. , and the sample is taken.
2. How the sample is.

- If any groups have been , it won't be random.
- If it isn't big enough, it won't be .
- Bigger samples should be more .

Estimating Population Size

Use a capture-recapture method:

1. Take a of a population, them and them.
2. Take a later on and record the that are tagged.
3. Assume the fraction of tagged members in the is the fraction of tagged members in the .

EXAMPLE

An ecologist catches, tags and releases 10 badgers in a forest. She returns 2 weeks later and catches 15 badgers. 2 of them are tagged. Work out an estimate for the population of badgers in the forest.

① → $\frac{\quad}{P} = \frac{\quad}{\quad}$ ← ②

③ so $P = \frac{\quad \times \quad}{\quad} = \quad$

Second Go:
...... /...... /......

Sampling and Data Collection

Definitions of Sampling Terms

	The whole group you want to find out about.
SAMPLE	
RANDOM SAMPLE	
	Fairly represents the whole population.
	Doesn't fairly represent the whole population.
QUALITATIVE DATA	
QUANTITATIVE DATA	
	Data that can only take exact values.
	Data that can take any value in a range.

Choosing a Simple Random Sample

1. Give
2. Make
3. Pick

Random numbers can be chosen using a

....

Spotting Bias

Two things to think about:

1. When
2. How

- If any groups
- If it isn't
- Bigger

Estimating Population Size

Use a capture-recapture method:

1. Take a random sample
2. Take a second random sample
3. Assume the fraction

EXAMPLE

An ecologist catches, tags and releases 10 badgers in a forest. She returns 2 weeks later and catches 15 badgers. 2 of them are tagged. Work out an estimate for the population of badgers in the forest.

 = 

③ so P = =

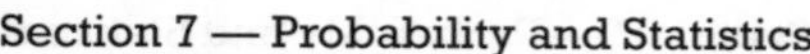

Averages and Ranges

First Go: ..... /..... /.....

Mean, Median, Mode and Range

MEAN	÷
	Middle value (when values are in size order)
MODE	
	Difference between highest and lowest values

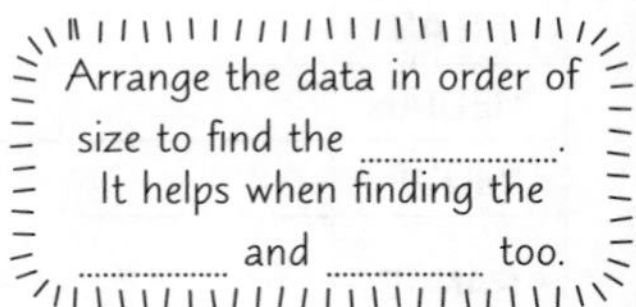

EXAMPLE

Find the mean, median, mode and range for the data below:
2.4 2.8 1.7 3.4 2.6 3.6 2.4 1.9

Mean = $\frac{\quad + \quad + \quad + \quad + \quad + \quad + \quad + \quad}{\quad} = \frac{\quad}{\quad} =$

The ....th value is halfway between the .....th and .....th value.

In order:

Median = th value = Mode =

Range = − =

Quartiles

Formulas are for a data set with n values.

LOWER QUARTILE, Q_1	The value (%) of the way through a data set	
	The value halfway (50%) through a data set	$\frac{n+1}{2}$
UPPER QUARTILE, Q_3	The value (%) of the way through a data set	
INTERQUARTILE RANGE, IQR	Difference between and (contains middle % of the data)	−

The range is affected by ...., the IQR is not.

Box Plots

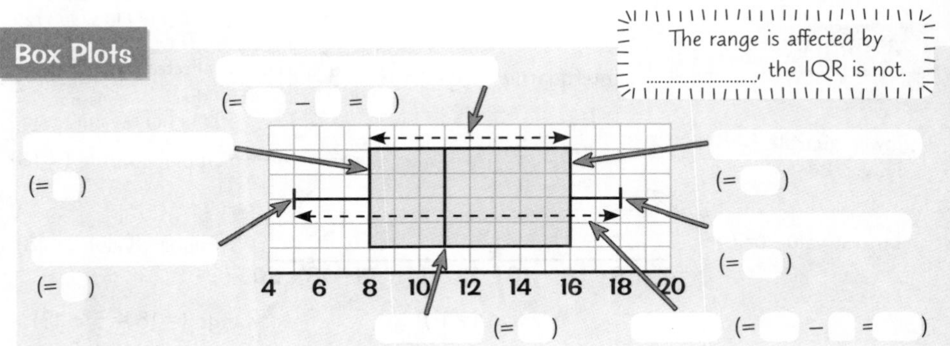

Second Go: ..... /..... /.....	**Averages and Ranges**

Averages and Ranges

Mean, Median, Mode and Range

MEAN	
MEDIAN	
MODE	
RANGE	

Arrange the data .... to find the .... It helps when finding the .... and .... too.

EXAMPLE

Find the mean, median, mode and range for the data below:

2.4 2.8 1.7 3.4 2.6 3.6 2.4 1.9

Mean = ———— = ——— =

In order:

Median = =   Mode =

Range = =

The .... value is .... the .... and .... value.

Quartiles

Formulas are for a data set with n values.

LOWER QUARTILE, Q_1		
MEDIAN, Q_2		
UPPER QUARTILE, Q_3		
INTERQUARTILE RANGE, IQR		

Box Plots

The .... is affected by outliers, the .... is not.

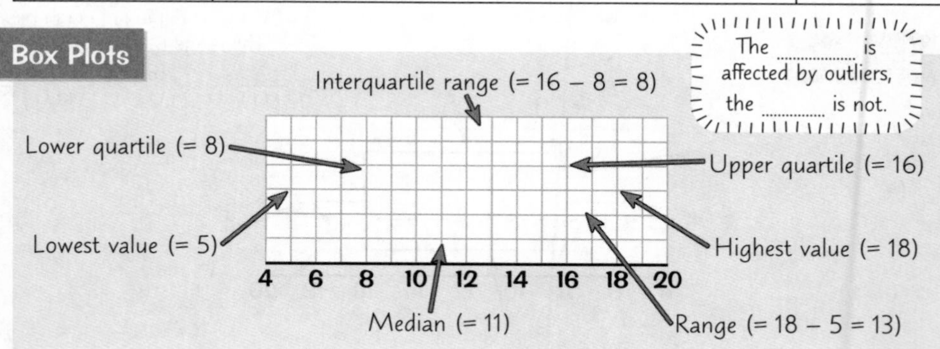

Frequency Tables

First Go: ..... /..... /.....

Finding Averages from Frequency Tables

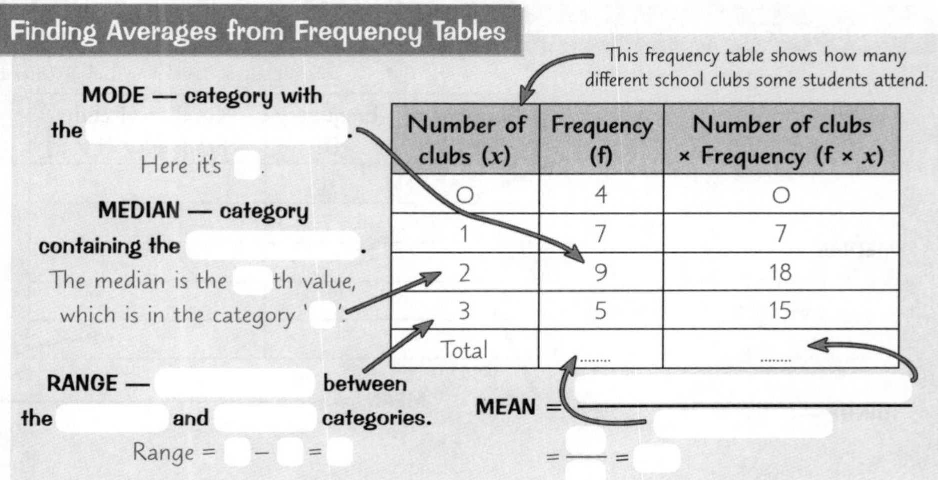

This frequency table shows how many different school clubs some students attend.

MODE — category with the ______.
Here it's ______.

MEDIAN — category containing the ______.
The median is the ______ th value, which is in the category '______'.

Number of clubs (x)	Frequency (f)	Number of clubs × Frequency (f × x)
0	4	0
1	7	7
2	9	18
3	5	15
Total	....	....

RANGE — ______ between the ______ and ______ categories.
Range = ______ – ______ = ______

MEAN = ______ / ______
= ______ / ______ = ______

Grouped Frequency Tables

Data is grouped into classes, with no gaps between classes for ______ data.

______ ______ are used to cover all possible values.

Height (h cm)	Frequency (f)	Mid-interval value (x)	f × x
0 < h ≤ 20	12	10	120
20 < h ≤ 30	28	....	700
30 < h ≤ 40	10	35	....
Total	....	—	....

Find the mid-interval value by adding up the ______ and dividing by ______.

MODAL CLASS — class with ______.
Here it's ______.

CLASS CONTAINING THE MEDIAN — contains the ______ of data.
Median is the ______ th value, so class containing the median is ______.

RANGE — ______ between the ______ and ______.
Estimated range = ______ – ______ = ______ cm

MEAN — multiply the ______ (x) by the ______ (f). Divide the ______ by the ______.
Estimated mean = ______ / ______ = ______ cm

You don't know the .... for grouped data so can only .... the mean and range.

Second Go: ...../...../.....

Frequency Tables

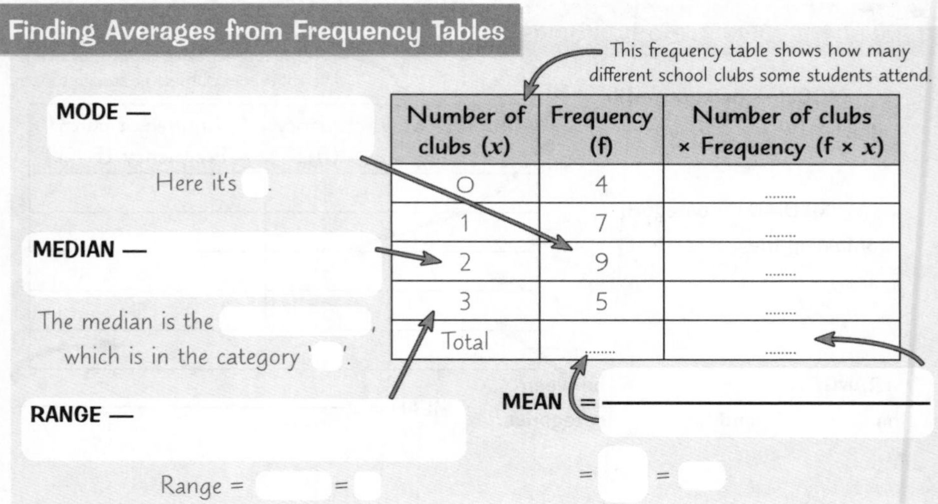

Finding Averages from Frequency Tables

This frequency table shows how many different school clubs some students attend.

Number of clubs (x)	Frequency (f)	Number of clubs × Frequency (f × x)
0	4	....
1	7	....
2	9	....
3	5	....
Total	....	....

MODE —

Here it's .

MEDIAN —

The median is the , which is in the category ' '.

RANGE —

Range = =

MEAN =

= =

Grouped Frequency Tables

Data is grouped into classes, with between classes for .

Inequality symbols are used to

Height (h cm)	Frequency (f)	Mid-interval value (x)	f × x
$0 < h \le 20$	12	....	....
$20 < h \le 30$	28	....	....
$30 < h \le 40$	10	....	....
Total	....	—	....

Find the mid-interval value by

MODAL CLASS — . Here it's .

CLASS CONTAINING THE MEDIAN — .

Median is the , so class containing the median is .

RANGE —

Estimated range = = cm

MEAN —

Estimated mean = = cm

You don't know the .... for .... so can only .... the .... and ....

Cumulative Frequency

First Go: ..... /..... /.....

Drawing Cumulative Frequency Graphs

CUMULATIVE FREQUENCY — the of the frequencies.

Width (w cm)	Freq.	Cumulative frequency
30 < w ≤ 40	8	8
40 < w ≤ 50	15	23
50 < w ≤ 60	26	....
60 < w ≤ 70	20	69
70 < w ≤ 80	11	....

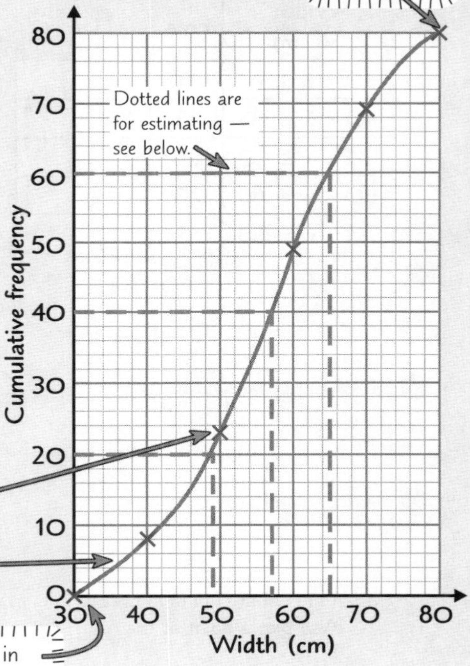

1. Add a column and the working down the table.
2. Plot the points — use the in each class and the .
3. with a smooth curve or straight lines.

Also plot a point at the .... in the ...., with cumulative frequency .....

Cumulative frequency (0–80) against Width (cm) (30–80)

Dotted lines are for estimating — see below.

.... of data values.

You can also .... percentiles — e.g. the 20th percentile is ....% of the way through the data.

Estimating From Cumulative Frequency Graphs

Go up to the value on the ,
across to the , then down and read off the .

- **To find the median, use the value through the .**
  In the example above, that's 40 — so the median is approximately .
- **For the lower and upper quartiles, use the values % and % of the way through.**
  Here, that's 20 and 60 — so $Q_1 \approx$ and $Q_3 \approx$. Then IQR ≈ – = .

To estimate the number of values less than or greater than a given value:

1. **Draw a line from that value on the to the curve.**
2. **Draw a line to read off the .**

Second Go: ...../...../.....

Cumulative Frequency

Drawing Cumulative Frequency Graphs

CUMULATIVE FREQUENCY — .

Width (w cm)	Freq.	Cumulative frequency
30 < w ≤ 40	8	....
40 < w ≤ 50	15	....
50 < w ≤ 60	26	....
60 < w ≤ 70	20	....
70 < w ≤ 80	11	....

1. Add a column and
2. Plot the points — use the
3. Join the points with a

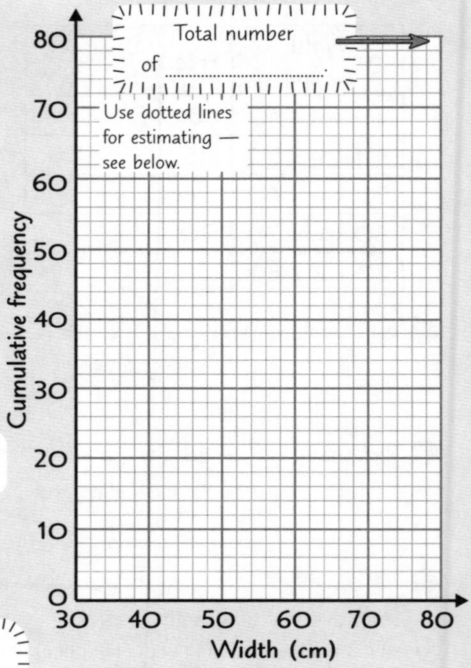

Also plot a point at the ...., with ....

Total number of ....

Use dotted lines for estimating — see below.

Cumulative frequency: 0, 10, 20, 30, 40, 50, 60, 70, 80

Width (cm): 30, 40, 50, 60, 70, 80

You can also estimate .... — e.g. the .... is 20% of the way through the data.

Estimating From Cumulative Frequency Graphs

Go up to ,

across , **then down** .

- **To find the median,** . In the example above, that's — so the median is approximately .
- **For the lower and upper quartiles,** .

  Here, that's and — so $Q_1 \approx$ and $Q_3 \approx$. Then IQR ≈ = .

To estimate the number of values less than or greater than a given value:

1. **Draw a line** .
2. **Draw a line** .

Histograms and Scatter Graphs

First Go: ...../...../.....

Histograms and Frequency Density

Two differences between histograms and bar charts:

1. The vertical axis of a histogram shows ______, not ______.
2. The bars on a histogram can be ______.

Frequency Density = ______ ÷ ______

Frequency = ______ × ______ = ______

Height (h cm)	Freq.	Frequency density
0 < h ≤ 20	16	0.8
20 < h ≤ 25	20	....
25 < h ≤ 40	30	....
40 < h ≤ 50	25	2.5

Add a ______ column.

Area = ... × .... = ... = Frequency

Use the second formula to estimate the frequency in .... of a class — just work out the area of that .... of the bar.

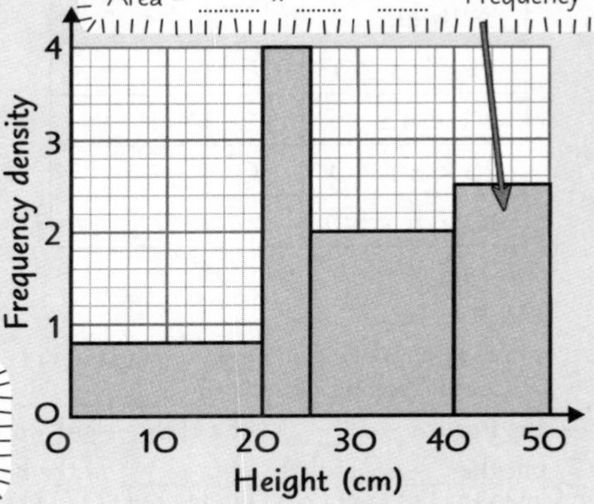

Scatter Graphs and Correlation

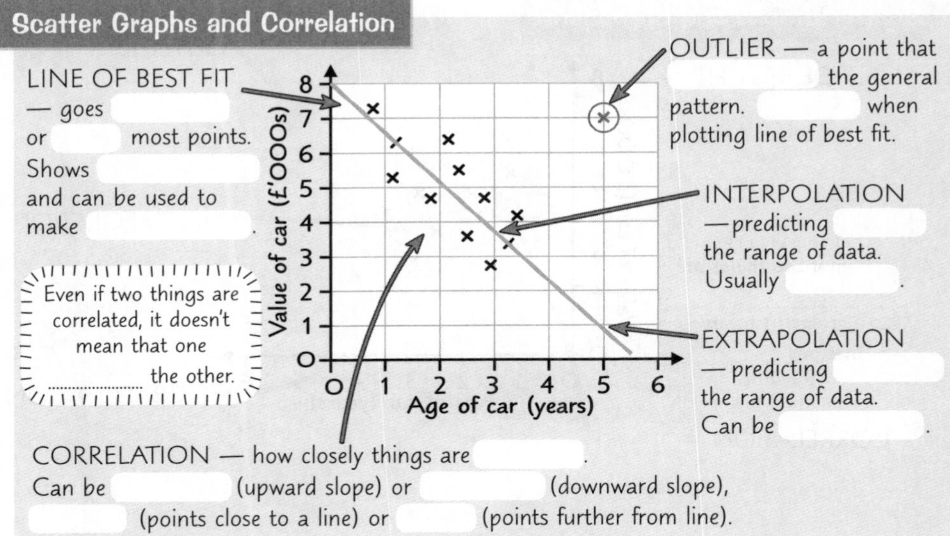

LINE OF BEST FIT — goes ______ or ______ most points. Shows ______ and can be used to make ______.

OUTLIER — a point that ______ the general pattern. ______ when plotting line of best fit.

INTERPOLATION — predicting ______ the range of data. Usually ______.

EXTRAPOLATION — predicting ______ the range of data. Can be ______.

Even if two things are correlated, it doesn't mean that one .... the other.

CORRELATION — how closely things are ______. Can be ______ (upward slope) or ______ (downward slope), ______ (points close to a line) or ______ (points further from line).

Second Go:
...../...../.....

Histograms and Scatter Graphs

Histograms and Frequency Density

Two differences between histograms and bar charts:

1 **The vertical axis**

2 **The bars**

Frequency Density =

Frequency = =

Height (h cm)	Freq.	Frequency density
0 < h ≤ 20	16	....
20 < h ≤ 25	20	....
25 < h ≤ 40	30	....
40 < h ≤ 50	25	....

Add a

Area = ... = ... = ...

Frequency density (0–4) vs Height (cm) (0–50)

Use the second formula to .... the frequency in .... of a class — just work out the .... of that .... of the bar.

Scatter Graphs and Correlation

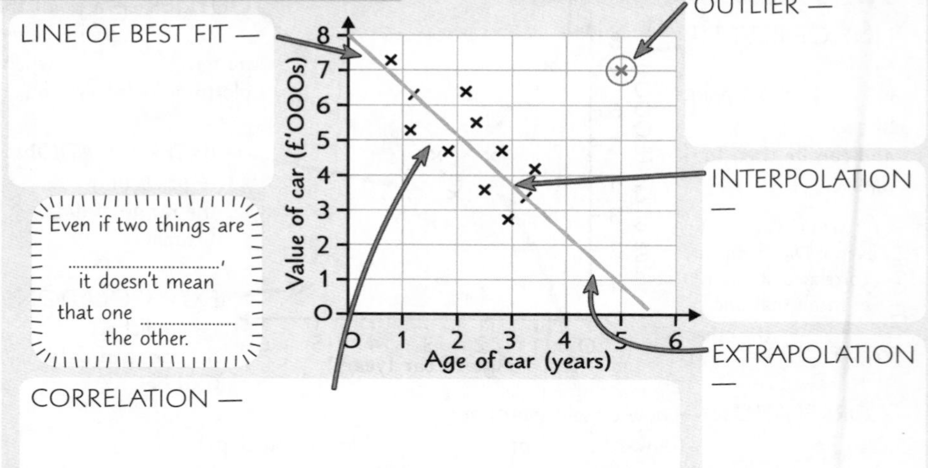

Even if two things are ...., it doesn't mean that one .... the other.

CORRELATION —

Other Graphs and Charts

First Go: ..... /..... /.....

Time Series

TIME SERIES — a line graph showing
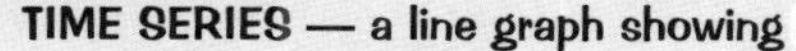
(a).

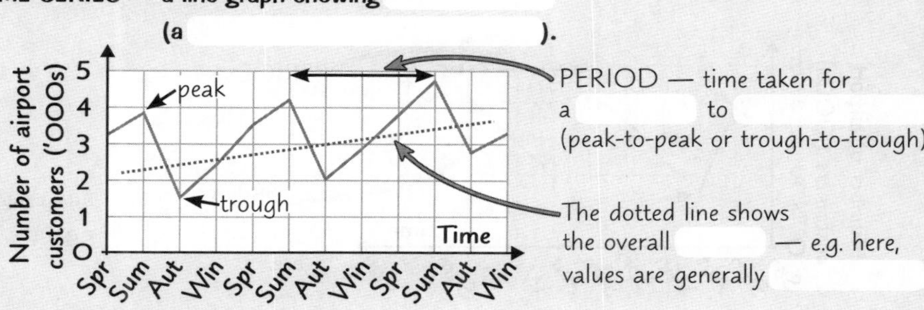

PERIOD — time taken for a to (peak-to-peak or trough-to-trough).

The dotted line shows the overall — e.g. here, values are generally .

Frequency Polygons

FREQUENCY POLYGON — displays from a .

is plotted against the

and points are joined with .

Weight (w kg)	Freq.
10 < w ≤ 12	9
12 < w ≤ 14	14
14 < w ≤ 16	18
16 < w ≤ 18	10

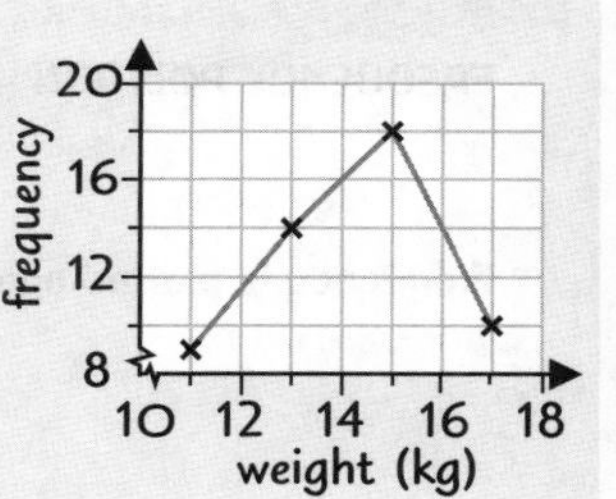

Pie Charts

PIE CHART — shows .

Total of all data =

This pie chart shows how 120 pupils travel to school:

The angle for the 'train' sector is

– – – =

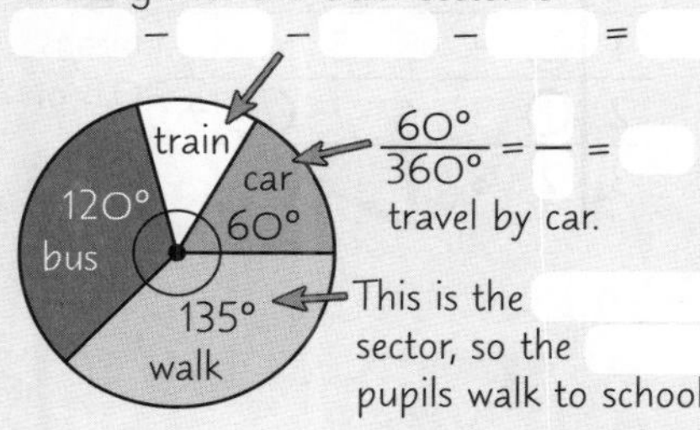

$\frac{60°}{360°}$ = — = travel by car.

This is the sector, so the pupils walk to school.

Each pupil is represented by $\frac{360°}{120}$ = .

Stem and Leaf Diagrams

STEM AND LEAF DIAGRAM — shows the of data.

Use them to find and ranges.

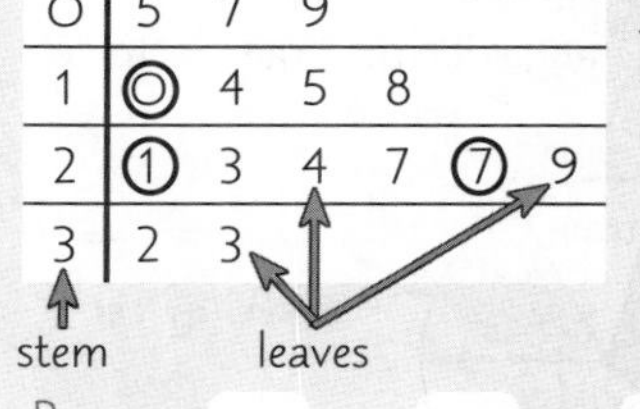

Key: 0 | 5 means 0.5 m

Range = m – m = m

Mode = m

Median = m

Q_1 = m, Q_3 = m

IQR = m – m = m

Second Go: ...../...../.....

Other Graphs and Charts

Time Series

TIME SERIES —

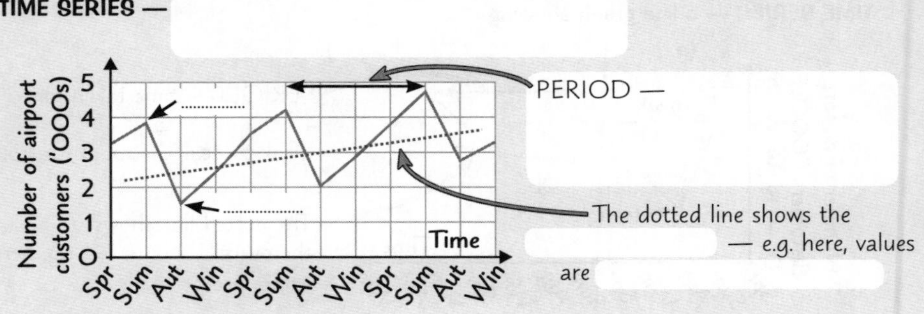

PERIOD —

The dotted line shows the — e.g. here, values are

Frequency Polygons

FREQUENCY POLYGON —

Frequency is plotted against

Weight (w kg)	Freq.
10 < w ≤ 12	9
12 < w ≤ 14	14
14 < w ≤ 16	18
16 < w ≤ 18	10

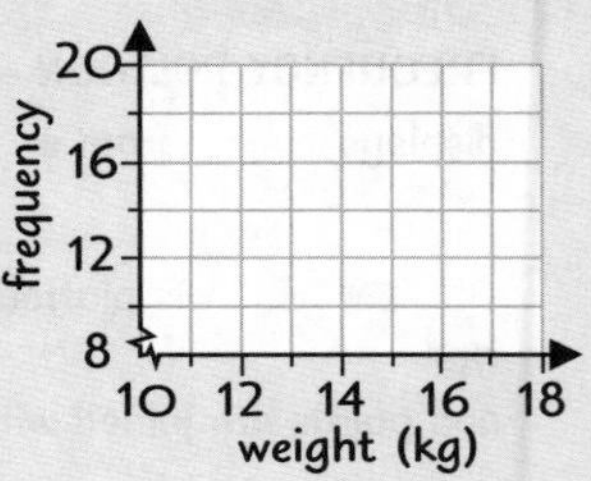

Pie Charts

PIE CHART —

Total of all data =

This pie chart shows how 120 pupils travel to school:

The angle for the 'train' sector is

=

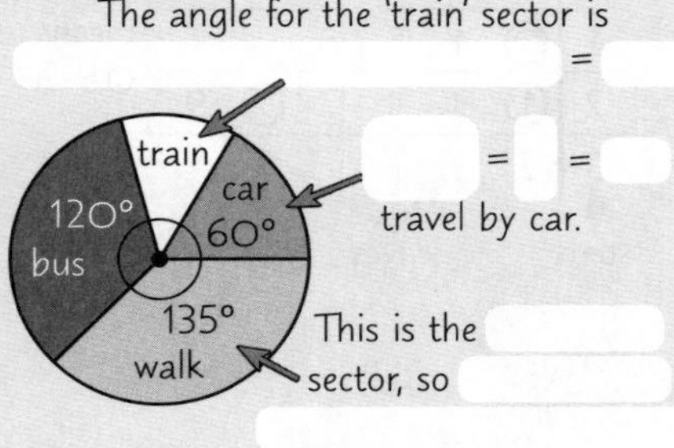

= = travel by car.

This is the sector, so

.

Each pupil is represented by = .

Stem and Leaf Diagrams

STEM AND LEAF DIAGRAM —

Use them

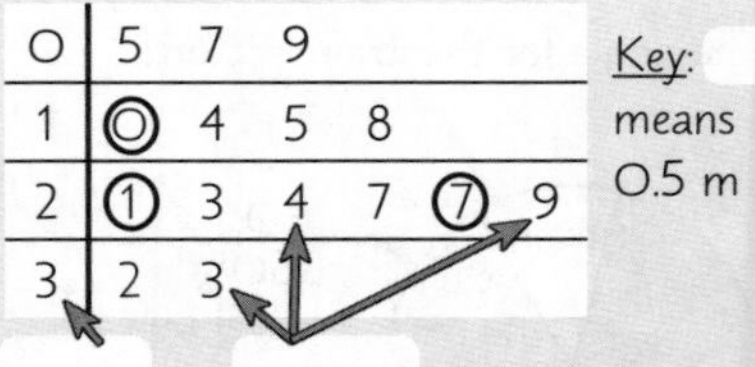

Key: means 0.5 m

Range = =

Mode = Median =

Q_1 = , Q_3 =

IQR = =

Mixed Practice Quizzes

Sincerest congratulations — you've made it to the final quizzes of the final section. One more set of quizzes on p.137-148, then you can put your feet up.

Quiz 1

Date: ...... / ...... / ......

1) What is the mode of a data set?
2) What do all the angles in a pie chart add up to?
3) What is correlation?
4) Why can you only estimate the mean for a grouped frequency table?
5) Which part of a box plot shows the median?
6) What is the formula for frequency density?
7) What is the formula to find the position of the upper quartile of a data set?
8) What is qualitative data?
9) What is meant by the 'period' on a time series?
10) Explain how you would estimate the lower quartile on a cumulative frequency graph.

Total:

Quiz 2

Date: ...... / ...... / ......

1) How do you find the mean from a frequency table?
2) What is the interquartile range?
3) How do you find the median of a data set?
4) What is the mid-interval value for the class $20 < x \leq 40$?
5) How far through a data set is the upper quartile?
6) What does the area of a bar on a histogram represent?
7) Describe weak negative correlation.
8) Explain how you would choose a simple random sample.
9) What is quantitative data?
10) For the entry 3 | 1 (where 3 | 1 means 31 cm), which number is the stem and which number is the leaf?

Total:

Mixed Practice Quizzes

Quiz 3

Date: ...... / ...... / ......

1) What is discrete data?
2) What is a line of best fit?
3) What is seasonality?
4) How do you find the angle that represents a single item when drawing a pie chart?
5) Does a box plot show the range, interquartile range, or both?
6) Describe two differences between a histogram and a bar chart.
7) For capture-recapture, what would you do after you'd caught, tagged and released a sample of the population?
8) What does strong positive correlation on a scatter graph look like?
9) Which two values do you use to estimate the range from a grouped frequency table?
10) When drawing a cumulative frequency curve, do you plot the lowest value in each class, the highest value in each class or the mid-interval value?

Total:

Quiz 4

Date: ...... / ...... / ......

1) How do you work out the mean of a data set?
2) What are stem and leaf diagrams useful for?
3) What is cumulative frequency?
4) Give one way in which a sample could be biased.
5) To find the mean from a grouped frequency table, what is the next thing you have to do once you've found the mid-interval values?
6) How do you find the median from a cumulative frequency graph?
7) How far through a data set is the lower quartile?
8) What is it called when you make predictions outside the range of your data?
9) What is a population?
10) What values do you plot on a frequency polygon?

Total:

Answers

Section 1 — Number

Pages 17-18

Quiz 1

Q1 Lower

Q2 Multiply together all of the prime factors that are in both numbers.

Q3 Put the digits after the decimal point as the numerator.

Q4 Turn the second fraction upside down and change the ÷ to ×.

Q5 Divide by the denominator and multiply by the numerator.

Q6 1, 4, 16

Q7 Brackets, Other, Division, Multiplication, Addition, Subtraction

Q8 True

Q9 No — the first number is smaller than 1.

Q10 A number whose only factors are 1 and itself.

Quiz 2

Q1 7

Q2 1 and 10

Q3 Positive

Q4 Irrational number

Q5 Prime numbers

Q6 $0.\dot{1}2\dot{6}$

Q7 Write 160 over 85 and simplify.

Q8 Rewrite one number so the powers of 10 are the same.

Q9 The lowest common multiple of the denominators.

Q10 Round all numbers to 1 or 2 s.f. then do the calculation with the rounded numbers.

Quiz 3

Q1 Divide by 100.

Q2 12.35

Q3 When a number is repeated.

Q4 Rearrange so the front numbers are together and powers of 10 are together.

Q5 A fraction with all nines as the denominator and the repeating part as the numerator.

Q6 Find a square number on each side of the given number.

Q7 The LCM is the smallest number that divides by all the numbers in question.

Q8 The multiplication

Q9 Make the denominators the same.

Q10 Write as an addition, turn the integer part into a fraction and add together.

Quiz 4

Q1 Negative

Q2 The lower bound of x and the upper bound of y

Q3 False — it's 2.5×10^{-3}.

Q4 2 and 5

Q5 **a)** 37.5% **b)** $\frac{3}{8}$

Q6 $2 \times 3^2 \times 5^2$

Q7 Rewrite them both as improper fractions.

Q8 10.8

Q9 Divide the numerator and denominator by the same number until they won't divide any more.

Q10 Multiply r by a power of 10 to get any non-repeating parts past the decimal point.

Section 2 — Algebra

Pages 27-28

Quiz 1

Q1 Multiply two brackets together, then multiply the result by the third bracket.

Q2 a^2b

Q3 False. $(x + \sqrt{y})^2 = x^2 + 2x\sqrt{y} + y$

Q4 $7x - 9y$

Q5 $3x$

Q6 Apply the power to numerator and denominator separately.

Q7 5

Q8 Multiply both sides by the denominator of the fraction.

Q9 Get the y^2 term on its own.

Q10 $\frac{1}{2}$

Quiz 2

Q1 $\sqrt{a} \times \sqrt{b} = \sqrt{a \times b}$

Q2 $24a^2$

Q3 Subtract the powers.

Q4 -1

Q5 Multiply the First terms in each bracket together, then the Outside terms, then Inside terms, then the Last terms, and collect like terms.

Q6 Square both sides of the formula.

Q7 1

Q8 The highest common factor of all terms in the expression.

Q9 $x = -3$ or $x = 3$

Q10 Flip it over and make the power positive.

Answers

Quiz 3

Q1 A collection of numbers, letters and brackets, multiplied or divided together.

Q2 $x = -4$

Q3 Multiply everything inside the bracket by the term outside the bracket.

Q4 $(4 + y)(4 - y)$

Q5 Multiply the powers together.

Q6 Multiply both sides by 5.

Q7 $\sqrt{a} \div \sqrt{b} = \sqrt{\frac{a}{b}}$

Q8 1

Q9 Multiplying to get rid of the surd in the denominator of a fraction.

Q10 Taking the square root.

Quiz 4

Q1 $-8x^3$

Q2 Add the powers together.

Q3 When an expression can be written in the form $a^2 - b^2$.

Q4 Multiply out both brackets.

Q5 Factorise to get the subject on its own.

Q6 $(xy)^3 = x^3y^3$

Q7 Multiply the numerator and denominator by $\sqrt{q}$.

Q8 It stays the same.

Q9 $x = \frac{y+11}{4}$

Q10 25

Pages 37-38

Quiz 1

Q1 $ax^2 + bx + c = 0$

Q2 29

Q3 $x = \frac{2}{3}$ and $x = -4$

Q4 You get two solutions — one when you add the square root, and one when you subtract.

Q5 -4

Q6 $x = -2 + \sqrt{5}$ and $x = -2 - \sqrt{5}$

Q7 Add the previous two terms together.

Q8 $(x + 1)(x + 5)$

Q9 Turn the second fraction upside down.

Q10 2

Quiz 2

Q1 Find the pair that add/subtract to give b.

Q2 Take out a factor of a from the first two terms.

Q3 The amount the sequence increases/decreases by each time.

Q4 Factorise the numerator and denominator and cancel any common factors.

Q5 Find the value that makes each bracket equal to 0.

Q6 Geometric sequence

Q7 Find a common denominator for the fractions.

Q8 -5

Q9 A sequence where the difference between terms changes each time. The nth term rule has an n^2 term.

Q10 $\left(-\frac{5}{2}, -\frac{11}{2}\right)$

Quiz 3

Q1 $4x^2 - 5x - 3 = 0$

Q2 Set the nth term rule equal to the number and solve — it's in the sequence if n is an integer.

Q3 Multiply top and bottom of the first fraction by $(x + 1)$, and top and bottom of the second fraction by x, to get them over a common denominator.

Q4 The quadratic doesn't factorise, the question mentions d.p. or s.f., you need exact answers/surds.

Q5 A sequence that increases or decreases by the same amount each time.

Q6 Writing a quadratic in the form $a(x + m)^2 + n$.

Q7 Multiply the numerators and denominators separately.

Q8 $a = 4$, $b = -7$, $c = 3$

Q9 Subtract the n^2 term from each term in the sequence. (This gives a linear sequence.)

Q10 $a = 3$, $b = -11$

Quiz 4

Q1 Writing it as two brackets multiplied together.

Q2 18

Q3 **a)** $\frac{y^2}{2x}$ **b)** $\frac{3}{x-1}$

Q4 $(x - 2)^2$

Q5 $x = \frac{-b \pm \sqrt{b^2 - 4ac}}{2a}$

Q6 $3n + 2$

Q7 The minimum y-value (if the coefficient of x^2 is positive), or maximum y-value (if the coefficient of x^2 is negative).

Q8 Yes

Q9 $(x + 1)(x - 1)$ or $x^2 - 1$

Q10 A sequence where you multiply/divide the previous term by the same number each time.

Pages 49-50

Quiz 1

Q1 Two

Q2 Something that takes an input, processes it and outputs a value.

Q3 $-2, -1, 0, 1$

Q4 Multiply out the brackets, collect like terms, then take out a factor of 4.

Q5 $x \geq a$ or $x \leq -a$

Q6 Substitute values into the equation until the sign changes.

Q7 **a)** $(x + 1)^2$ **b)** $x^2 + 1$

Q8 E.g. $2n + 1$

Q9 Dotted

Q10 The inverse function of g(x).

Quiz 2

Q1 $x \geq 3$

Q2 Substitute the values into the equation you didn't use to find the solution, and check they both work.

Q3 Proving something is false by finding an example that doesn't work.

Q4 Substitute a number in place of x.

Q5 Flip it to point the other way.

Q6 When it's too hard to solve.

Answers

Q7 Write expressions for two odd numbers, add them, and show it can be written as 2 × something.

Q8 $y = 3x - 2$ (with a dotted line) and $y = 6 - 2x$ (with a solid line).

Q9 $f^{-1}(x) = \frac{x-2}{5}$

Q10 The nth value in the iteration (e.g. x_2 is the second value).

Quiz 3

Q1 $f(x) = 3x + 7$

Q2 Hollow

Q3 E.g. $2n$

Q4 A method where a calculation is repeated to get closer to the actual solution.

Q5 Replace every x in $f(x)$ with the expression for $g(x)$.

Q6 Rearrange one equation so a non-quadratic unknown is by itself.

Q7 A function that reverses a given function.

Q8 $-6 < x < 6$

Q9 1.2

Q10 E.g. $2 + 3 = 5$

Quiz 4

Q1 E.g. n, $n + 1$

Q2 $x > -7$

Q3 **a)** $f(2) = -3$ **b)** $f(-1) = 9$

Q4 A set of instructions to find a solution to an equation to a certain degree of accuracy.

Q5 Multiply both equations so the coefficients of either x or y are the same.

Q6 It means the statement is always true (i.e. it works for all values).

Q7 Two functions combined into a single function.

Q8 Substitute a point into the inequality. If the inequality is true, that point is on the correct side of the line.

Q9 $(0)^4 + 9(0) - 7 = -7$ and $(1)^4 + 9(1) - 7 = 3$. Sign change means there is a solution.

Q10 Substitute a number in, and see if the inverse reverses the function.

Section 3 — Graphs

Pages 59-60

Quiz 1

Q1 $y = ax^3 + bx^2 + cx + d$

Q2 False — perpendicular lines have gradients that multiply to give -1.

Q3 A vertical line that passes through 2 on the x-axis.

Q4 n-shaped

Q5 Add the x-coordinates of the end points and divide by 2. Then add the y-coordinates of the end points and divide by 2.

Q6 2

Q7 $y = -2x + 5$

Q8 Substitute the x-values into the equation to get the y-values. Plot the points and join with a smooth curve.

Q9 $(1, -2)$

Q10 $x = 0$

Quiz 2

Q1 -4

Q2 $y = 6$

Q3 $y = 2x + 3$

Q4 $(-2, 0)$ and $(3, 0)$

Q5 $(0, 0)$

Q6 True

Q7 Draw a table with three values of x. Put the x-values into the equation and work out the y-values. Plot the points and draw a line through them.

Q8 True

Q9 A horizontal line

Q10 A right angle

Quiz 3

Q1 $y = 4x + 3$

Q2 $y = ax^2 + bx + c$

Q3 $\text{gradient} = \frac{\text{change in } y}{\text{change in } x}$

Q4 A quadratic

Q5 $(0, -8)$ and $(2, 0)$

Q6 The gradient

Q7 True

Q8 $y = x$

Q9 The graph has a wiggle in the middle and goes down from the top left.

Q10 -3

Quiz 4

Q1 $y = 6x - 9$

Q2 A cubic

Q3 $(4, 6)$

Q4 Set $x = 0$ and find y. Set $y = 0$ and find x. Mark and label both points. Draw a line through them.

Q5 -1

Q6 c

Q7 False — the equation of the x-axis is $y = 0$.

Q8 They will have the same gradient (m).

Q9 Use both points to find the gradient. Substitute one point into $y = mx + c$. Rearrange to find c. Write the equation as $y = mx + c$.

Q10 $p = 3$

Pages 69-70

Quiz 1

Q1 2

Q2 A translation of 8 units up.

Q3 $y = 1$

Q4 Speed

Q5 True

Q6 Join the points with a straight line and find the gradient of the straight line.

Q7 $y = 4x - 2$

Q8 Divide the area into trapeziums.

Q9 The y-axis

Q10 False — the graph of $\tan x$ passes through $(0, 0)$ so must be defined for 0°.

Answers

Quiz 2

Q1 $y = x$ and $y = -x$

Q2 $y = -8x + 2$

Q3 Draw the graphs of both equations and find the coordinates where the graphs cross.

Q4 A translation of 2 units left.

Q5 Draw a tangent to the curve at the point and find the gradient of the tangent.

Q6 That the object has stopped.

Q7 (0, 1)

Q8 False — they repeat every 360°.

Q9 That the object is decelerating.

Q10 +1 and –1

Quiz 3

Q1 $y = k^x$ or $y = k^{-x}$

Q2 f(x) + a

Q3 False — it shows that the object is moving at a steady speed.

Q4 Upwards

Q5 180°

Q6 The top right and bottom left quadrants.

Q7 $x^2 + y^2 = 100$

Q8 $y = (x + 5)^2$

Q9 True

Q10 metres per second (m/s)

Quiz 4

Q1 f(x) – 3 is the graph of f(x) translated 3 units down.

Q2 That the object is coming back.

Q3 $x = 0$

Q4 The total distance travelled.

Q5 The general shape of an exponential graph with $0 < k < 1$ is flipped horizontally compared to an exponential graph with $k > 1$.

Q6 $y = \cos x$

Q7 Add 2 to the end of the equation.

Q8 True

Q9 $y = \tan x$

Q10 A reciprocal graph

Section 4 — Ratio, Proportion and Rates of Change

Pages 83-84

Quiz 1

Q1 Write one number on top of the other.

Q2 Percentage change $= \frac{\text{change}}{\text{original}} \times 100$

Q3 The other quantity is also tripled.

Q4 Divide to find the value of one part.

Q5 Multiply by 100 then by 100 again (or multiply by 10 000).

Q6 1.075

Q7 Convert to the smaller unit, then divide all numbers by the same thing.

Q8 In simple interest, the same amount (a percentage of the original amount) is added each time. In compound interest, the amount changes (it's a percentage of the new amount).

Q9 False — parts in a ratio are always in direct proportion.

Q10 Divide £230 000 by 115 to find the value of 1%.

Quiz 2

Q1 A direct proportion graph is a straight line through the origin, increasing from left to right.

Q2 Turn the percentage into a fraction or decimal then multiply by the amount.

Q3 Add the parts together to find the denominator, then write the part you want as the numerator.

Q4 Write the amount as a percentage of the original value.

Q5 Subtract the part from the whole.

Q6 $N = N_0 \times (\text{multiplier})^n$

Q7 Multiply to get rid of the fractions, then divide all numbers by the same thing.

Q8 $y \propto \frac{1}{x^2}$

Q9 1000 cm^3

Q10 speed $= \frac{\text{distance}}{\text{time}}$

Quiz 3

Q1 The other quantity doubles.

Q2 Convert 0.8 litres to ml.

Q3 Multiply the original value by the multiplier.

Q4 Find 2.5% of the original value and multiply it by 5.

Q5 1000 kg

Q6 Work out what the side has been multiplied by to get to the actual value, then multiply the other side(s) by this value.

Q7 Density $= \frac{\text{mass}}{\text{volume}}$

Q8 Add to find the total number of parts.

Q9 **a)** 1000 **b)** 1.015 **c)** 4

Q10 $y = k\sqrt[3]{x}$

Quiz 4

Q1 Divide the first number by the second then multiply by 100.

Q2 Pressure $= \frac{\text{force}}{\text{area}}$

Q3 An inverse proportion graph is a reciprocal graph.

Q4 k = 4

Q5 Multiply both sides by 10 to get rid of the decimals.

Q6 0.86

Q7 17

Q8 The initial amount

Q9 Divide by 10, then by 10, then by 10 again (or divide by 1000).

Q10 80 ml

Answers

Section 5 — Geometry and Measures

Pages 91-92

Quiz 1

Q1 70°

Q2 100°, 80° and 80°

Q3 360°

Q4 80°

Q5 1

Q6 True

Q7 Isosceles triangle

Q8 7

Q9 Order 2

Q10 False — opposite angles in a cyclic quadrilateral add up to 180°.

Quiz 2

Q1 True

Q2 90°

Q3 Equilateral triangle

Q4 60°

Q5 90°

Q6 180°

Q7 The angle in the opposite segment (by the alternate segment theorem).

Q8 180°

Q9 $(n-2) \times 180°$

Q10 6 cm

Quiz 3

Q1 2

Q2 True — this radius is perpendicular to the chord and passes through the circle's centre, so it bisects the chord.

Q3 Kite

Q4 Alternate angles

Q5 False — a trapezium has no rotational symmetry.

Q6 135°

Q7 The angle between a tangent and a chord is equal to the angle in the opposite segment.

Q8 150°

Q9 Nonagon

Q10 Twice as big.

Quiz 4

Q1 Square

Q2 90°

Q3 180°

Q4 90°

Q5 50°

Q6 180°

Q7 True

Q8 A regular polygon's number of sides is equal to the order of rotational symmetry.

Q9 3 cm

Q10 60°

Pages 99-100

Quiz 1

Q1 Major segment

Q2 False — the angle must be between the two sides for the triangles to be congruent.

Q3 The shape goes to the other side of the centre of enlargement.

Q4 The area of a circle.

Q5 (–2, 3)

Q6 The right angle, hypotenuse and another side all match up.

Q7 Proportional

Q8 Scale factor = $\frac{\text{new length}}{\text{old length}}$

Q9 3 cm^2

Q10 The angle, direction and centre of rotation.

Quiz 2

Q1 It goes to the other side of the centre of enlargement and doubles in size.

Q2 Minor sector

Q3 Enlargement

Q4 Split it into triangles and quadrilaterals, work out each area separately and add them up.

Q5 Arc length $= \frac{x}{360} \times$ circumference

Q6 SSS, ASA, SAS, RHS

Q7 The bottom number.

Q8 **a)** Yes **b)** No

Q9 The direction of rotation.

Q10 Area = base × vertical height

Quiz 3

Q1 The scale factor and the centre of enlargement.

Q2 Congruent

Q3 $\begin{pmatrix} 3 \\ -2 \end{pmatrix}$

Q4 Sector area $= \frac{x}{360} \times$ area of circle

Q5 3

Q6 –1

Q7 The outside edges.

Q8 True

Q9 Area $= \frac{1}{2} \times$ base × vertical height

Q10 3

Quiz 4

Q1 The equation of the mirror line.

Q2 True

Q3 Area $= \frac{1}{2}(a+b) \times$ vertical height

Q4 Find the area of the minor sector and subtract the area of the triangle made by the radii and the chord.

Q5 7.5 cm

Q6 The dimensions of the shape become a quarter of their original size.

Q7 Circumference $= \pi D$ or $2\pi r$

Q8 That a corresponding side in each triangle matches up.

Q9 (12, 1)

Q10 Major sector

Answers

Pages 111-112

Quiz 1

Q1 False — there is only one triangle you could draw.

Q2 Circle

Q3 8 vertices, 12 edges and 6 faces.

Q4 Increase them when drawing the second set of arcs.

Q5 Subtract the volume of the removed cone from the volume of the full cone.

Q6 A line perpendicular to the given line and a line perpendicular to that new line.

Q7 Clockwise

Q8 n^2

Q9 Volume = cross-sectional area × length

Q10 Measure towards the dot and label the point.

Quiz 2

Q1 1:8

Q2 The locus bisects the angle.

Q3 Surface area = $\pi rl + \pi r^2$

Q4 Plan

Q5 How fast the volume is changing.

Q6 3

Q7 090°

Q8 Find the area of its net.

Q9 A line or region showing all points that fit a given rule.

Q10 Roughly sketch and label the triangle.

Quiz 3

Q1 The locus of points equidistant from points A and B.

Q2 6 vertices, 9 edges and 5 faces.

Q3 Label all known points and sides.

Q4 8 cm^3

Q5 Multiply the map distance by the scale factor.

Q6 The surface area of a cylinder.

Q7 Draw lines from the ends through the dots and label the intersection.

Q8 From point B.

Q9 Volume = $\frac{4}{3}\pi r^3$

Q10 Keep them the same.

Quiz 4

Q1 Draw arcs on the lines from the intersection, then draw another arc from each of the first arcs.

Q2 5 km

Q3 The locus of points at a fixed distance from a given line.

Q4 Slant height

Q5 Set compasses to each side length and draw an arc from each end.

Q6 The surface area of a sphere.

Q7 060°

Q8 From the given point.

Q9 Volume = $\frac{1}{3}\pi r^2 h_v$

Q10 2

Section 6 — Pythagoras and Trigonometry

Pages 125-126

Quiz 1

Q1 Hypotenuse

Q2 It changes size.

Q3 $\frac{1}{\sqrt{2}}$

Q4 $\tan x = \frac{\text{opp}}{\text{adj}}$

Q5 $\frac{a}{\sin A} = \frac{b}{\sin B} = \frac{c}{\sin C}$

Q6 To find a side length.

Q7 Label the sides O, A and H.

Q8 $\sqrt{3}$ cm^2

Q9 5 cm

Q10 Show that $\overrightarrow{PQ}$ is a scalar multiple of $\overrightarrow{QR}$ or $\overrightarrow{PR}$.

Quiz 2

Q1 $\sin x = \frac{\text{opp}}{\text{hyp}}$, $\cos x = \frac{\text{adj}}{\text{hyp}}$, $\tan x = \frac{\text{opp}}{\text{adj}}$

Q2 Top: horizontal distance moved. Bottom: vertical distance moved.

Q3 Area = $\frac{1}{2}ab \sin C$

Q4 $a^2 + b^2 = c^2$

Q5 tan 60°

Q6 $\sin x = \frac{\text{opp}}{\text{hyp}}$

Q7 Take the square root.

Q8 Sine rule

Q9 $\overrightarrow{QR} = \frac{5}{7}\underline{a}$

Q10 $\sqrt{83}$ cm

Answers

Quiz 3

Q1 $\frac{1}{2}$

Q2 Cosine rule

Q3 $a^2 = c^2 - b^2$ (or $a = \sqrt{c^2 - b^2}$)

Q4 Take the inverse.

Q5 $a^2 + b^2 + c^2 = d^2$ (or $d = \sqrt{a^2 + b^2 + c^2}$)

Q6 False — scalar multiples of vectors are parallel.

Q7 A / C × H (formula triangle: A on top, C × H on the bottom)

Q8 Form a triangle using the points, subtract the coordinates to find the shorter lengths, then use Pythagoras to find the hypotenuse.

Q9 $\cos A = \frac{b^2 + c^2 - a^2}{2bc}$

Q10 $\overrightarrow{AC} = \underline{a} - \underline{b}$

Quiz 4

Q1 12 cm

Q2 It changes size and reverses direction.

Q3 1

Q4 True

Q5 $a^2 = b^2 + c^2 - 2bc \cos A$

Q6 1, $\sqrt{3}$ and 2

Q7 $\cos x = \frac{\text{adj}}{\text{hyp}}$

Q8 Draw a right-angled triangle between the line and the plane.

Q9 $\begin{pmatrix} 2 \\ 8 \end{pmatrix}$

Q10 Two sides and the angle enclosed by them.

Section 7 — Probability and Statistics

Pages 135-136

Quiz 1

Q1 Impossible

Q2 0.08

Q3 For independent events, one event happening has no effect on the probability of the other event happening. For dependent events, one event happening affects the probability of the other event happening.

Q4 The total number of things.

Q5 P(A or B) = P(A) + P(B)

Q6 Expected frequency = probability × number of trials

Q7 Multiply along the branches.

Q8 The probability of event A happening given that event B happens.

Q9 Everything inside both circles.

Q10 0.55

Quiz 2

Q1 A diagram that shows all possible outcomes.

Q2 A collection of elements, written in curly brackets.

Q3 Probability $= \frac{\text{Number of ways for something to happen}}{\text{Total number of possible outcomes}}$

Q4 1

Q5 A ∩ B — the intersection of sets A and B.

Q6 15 times

Q7 Repeat the experiment.

Q8 Every outcome is equally likely.

Q9 16 outcomes

Q10 P(A and B) = P(A) × P(B given A)

Quiz 3

Q1 1

Q2 P(A or B) = P(A) + P(B) – P(A and B)

Q3 Certain

Q4 Relative frequency $= \frac{\text{Frequency}}{\text{Number of times you tried the experiment}}$

Q5 P(A|B)

Q6 $\frac{4}{9}$

Q7 0.45

Q8 To record results when experiments have more than one step.

Q9 They change.

Q10 8

Quiz 4

Q1 The number of elements in set A.

Q2 P(A and B) = P(A) × P(B)

Q3 How many times you'd expect something to happen in a certain number of trials.

Q4 The intersection of the circles for A and B.

Q5 Likely

Q6 Subtract the probability that the event happens from 1.

Q7 Some outcomes are more likely than others.

Q8 0.24

Q9 Dependent

Q10 Everything inside the rectangle.

Answers

Pages 149-150

Quiz 1

Q1 The most common value.

Q2 360°

Q3 How closely two things are related.

Q4 Because you don't know the actual values.

Q5 The vertical line inside the box.

Q6 Frequency density = frequency ÷ class width

Q7 Upper quartile = $\frac{3(n+1)}{4}$

Q8 Data described in words

Q9 The time taken for a pattern to repeat itself.

Q10 Draw a horizontal line one quarter of the way through the cumulative frequency total to the curve, then go down and read off the value from the bottom axis.

Quiz 2

Q1 Multiply the number of each category by the frequency, then add up these values. Then use the formula: mean $= \frac{\text{total (category} \times \text{frequency)}}{\text{total frequency}}$

Q2 The difference between the upper and lower quartiles.

Q3 Arrange the data in size order. The median is the middle value.

Q4 30

Q5 75% (or $\frac{3}{4}$)

Q6 Frequency

Q7 Points form a downward slope and are spread out from the line of best fit.

Q8 Give each member of the population a number. Make a list of random numbers. Pick the members of the population with those numbers.

Q9 Data described using numbers.

Q10 3 is the stem and 1 is the leaf.

Quiz 3

Q1 Data that can only take exact values.

Q2 A line that goes through or near most points on a scatter graph.

Q3 A basic repeating pattern on a time series.

Q4 Divide the total number of items by 360.

Q5 Both

Q6 The vertical axis on a histogram shows frequency density, but on a bar chart, it shows frequency. The bars on a histogram can be different widths, but on a bar chart, each bar is the same width.

Q7 Take a second sample later on and record the fraction that are tagged.

Q8 Points form an upward slope and are close to the line of best fit.

Q9 The highest and lowest class boundaries.

Q10 The highest value in each class.

Quiz 4

Q1 Add up the values and divide by the number of values.

Q2 Finding averages and ranges.

Q3 The running total of the frequencies.

Q4 E.g. Some groups could have been excluded. / It might not be big enough.

Q5 Multiply the mid-interval values by the frequency for each class.

Q6 Draw a horizontal line halfway through the cumulative frequency total to the curve, then go down and read off the value from the bottom axis.

Q7 25% (or $\frac{1}{4}$)

Q8 Extrapolation

Q9 The whole group you want to find out about.

Q10 Frequency against mid-interval value.